THEORY OF THERMOMECHANICAL PROCESSES IN WELDING

Theory of Thermomechanical Processes in Welding

by

Andrzej Służalec

Technical University of Czestochowa, Poland

Springer

A C.I.P. Catalogue record for this book is available from the Library of Congress.

ISBN 978-90-481-6762-3 (PB)
ISBN 979-1-4020-2991-2 (e-book)

Published by Springer,
P.O. Box 17, 3300 AA Dordrecht, The Netherlands.

Printed on acid-free paper

springeronline.com

Printed in the Netherlands.

Contents

Part III HEAT FLOW IN WELDING

Part IV WELDING STRESSES AND DEFORMATIONS

Preface

The main purpose of this book is to provide a unified and systematic continuum approach to engineers and applied physicists working on models of deformable welding material. The key concept is to consider the welding material as an thermodynamic system.

Significant achievements include thermodynamics, plasticity, fluid flow and numerical methods.

Having chosen point of view, this work does not intend to reunite all the information on the welding thermomechanics. The attention is focused on the deformation of welding material and its coupling with thermal effects.

Welding is the process where the interrelation of temperature and deformation appears throughout the influence of thermal field on material properties and modification of the extent of plastic zones. Thermal effects can be studied with coupled or uncoupled theories of thermomechanical response. A majority of welding problems can be satisfactorily studied within an uncoupled theory. In such an approach the temperature enters the stress-strain relation through the thermal dilatation and influences the material constants. The heat conduction equation and the relations governing the stress field are considered separately.

In welding a material is either in solid or in solid and liquid states. The flow of metal and solidification phenomena make the welding process very complex. The automobile, aircraft, nuclear and ship industries are experiencing a rapidly-growing need for tools to handle welding problems. The effective solutions of complex problems in welding became possible in the last two decades, because of the vigorous development of numerical methods for thermal and mechanical analysis.

The book has been divided into four parts. Part I Fundamentals of Welding Thermomechanics is devoted to the description of the deformation of the welding material. Both the Lagrangian and the Eulerian standpoints are considered. The concept of stress is introduced. The derivations of the theorem of virtual work rate and the kinetic energy theorem are provided. A general thermodynamic framework for the formulation of all constitutive equations is given. In this framework the laws governing the heat flow are formulated. The background of plasticity in welding is introduced. Hardening and softening phenomena are refined in the global context of thermodynamics and the modeling of thermal hardening in welding is discussed. In Part II Numerical Solutions of Welding Problems all the constitutive models of the previous chapters are discussed extending well-established standard procedures for solid and liquid media. The numerical approach to thermo-elastic-plastic material model is discussed. The phase-change problems are analyzed. The numerical solution of fluid flow in welding is presented. Part III Heat Flow in Welding presents in details the solution of heat transfer equations under conditions of interest in welding. The restrictive assumptions will however limit the practical utility of the results. Nevertheless the results are useful in that they emphasize the variables involved and approximate the way in which they are related. In addition, such solutions provide the background for understanding more complicated solutions obtained numerically and provide guidance in making judgements. The numerical solutions of welding problems are focused on laser and electroslag welding. In Part IV Welding Stresses and Deformations coupled thermomechanical phenomena are discussed. First of all thermal stresses being the result of complex temperature changes and plastic strains in regions near the weld are analyzed. Deformations and residual stresses in welding structures are considered for chosen welding processes and geometry of elements.

Częstochowa
May 2004

AndrzejS łużalec

Part I

FUNDAMENTALS OF WELDING THERMOMECHANICS

Chapter 1

DESCRIPTION OF WELDING DEFORMATIONS

1.1 Introduction

A material in the welding process can be considered by way of the concept of a body. In the course of thermal and mechanical loadings the body changes its geometrical shape. The consequences of welding processes are small deformations of the welded body and large deformations as it is in friction and spot welding. The deformation and the motion of welding material is described as standard continuum.

1.2 The Referential and Spatial Description

Consider a body B_t being a set of elements with the points of a region B of a Euclidean point space at current time t. The elements of B_t are called particles. A point in B is said to be occupied by a particle of B_t if a given particle of B_t corresponds with the point.

The position of any particle X is located by the position **X** of the point, relative to the origin. The components of **X** are called the referential coordinates of X. The terms Lagrangian and material are also used to describe these coordinates. After deformation of the body, the material is in a new configuration, called current configuration. In this configuration the points of the Euclidean point space are identified by their position vector **x**.

In a motion of a body, the configuration changes with the time t. The terms Eulerian or spatial coordinates are used to describe the motion. In a motion of a body B_t, a representative particle X occupies a succession of points which together form a curve in Euclidean point space.

The motion of the body is given as a function of the position in the reference configuration

$$\mathbf{x} = \mathbf{x}(\mathbf{X},t) \tag{1.2.1}$$

The referential position $\mathbf{X}$ and the time t defined on a reference configuration are the independent variables and the fields are said to be given in the referential description. The independent variables are $\mathbf{x}$ and t and the fields defined on the configurations constituting a motion of B_t are said to be given in the spatial description. A system of general referential and spatial coordinates in space is setup by adjoining to the respective origins *0* and *o* and the bases $\mathbf{G}_\alpha$ ($\alpha = 1,2,3$) and $\mathbf{g}_i$ ($i = 1,2,3$) in referential and spatial descriptions respectively. Referential coordinates are denoted by X_α ($\alpha = 1,2,3$) and spatial by x_i ($i = 1,2,3$).
Eq. (1.2.1) may be expressed in the form

$$x_k = x_k(X_\alpha,t) \tag{1.2.2}$$

A vector field $\mathbf{u}$ and any tensor field $\mathbf{A}$ have the form

$$\mathbf{u} = u_k\,\mathbf{g}_k = u_\mu\,\mathbf{G}_\mu \tag{1.2.3}$$

$$\mathbf{A} = A_{kl}\,\mathbf{g}_k \otimes \mathbf{g}_l = A_{\mu v}\,\mathbf{G}_\mu \otimes \mathbf{G}_v = A_{k\mu}\,\mathbf{g}_\mu \otimes \mathbf{G}_\mu \tag{1.2.4}$$

where the symbol $\otimes$ stands for the tensorial product.

1.3 The Deformation Gradient

A deformation is the mapping of a reference configuration into a current configuration. The fundamental kinematic tensor introduced in modern continuum mechanics is the deformation gradient $\mathbf{F}$ defined by

$$\mathbf{F} = \mathrm{Grad}\,\mathbf{x} \tag{1.3.1}$$

which can be expressed in component form as

$$\mathbf{F} = F_{\rho\mu}\,\mathbf{g}_\rho \otimes \mathbf{G}_\mu \tag{1.3.2}$$

$$F_{i\alpha} = x_{i,\alpha} \tag{1.3.3}$$

In Eq. (1.3.1) Grad denotes the gradient operator with regard to position in the reference configuration. Since Eq. (1.2.1) has an inverse and the mapping has continuous derivatives, it implies that **F** has an inverse $\mathbf{F}^{-1}$ defined by

$$\mathbf{F}^{-1} = \text{grad}\,\mathbf{X} \tag{1.3.4}$$

or in the component form

$$\mathbf{F}^{-1} = (\mathrm{F}^{-1})_{\mu\rho}\,\mathbf{G}_{\mu} \otimes \mathbf{g}_{\rho} \tag{1.3.5}$$

where

$$(\mathrm{F}^{-1})_{\mu\rho} = \mathrm{X}_{\alpha,\rho} \tag{1.3.6}$$

In Eq. (1.3.4) grad denotes the gradient operator with regard to position in the current configuration of the body B_t.
Eq. (1.3.1) can be expressed in the equivalent form

$$\mathrm{d}\mathbf{x} = \mathbf{F}\,\mathrm{d}\mathbf{X} \tag{1.3.7}$$

The Jacobian of the mapping **F** is

$$\mathrm{J} = \det \mathbf{F} \neq 0 \tag{1.3.8}$$

Consider unit tangent vectors **N** and **n** to any material curve in the current and reference configurations respectively, then

$$\mathrm{d}\mathbf{x} = \mathbf{n}\,\mathrm{da} \qquad \mathrm{d}\mathbf{X} = \mathbf{N}\,\mathrm{dA} \tag{1.3.9}$$

where da and dA are infinitesimal surfaces in current and reference configurations.
By (1.3.9) and (1.3.7) we get

$$\left(\frac{\mathrm{da}}{\mathrm{dA}}\right) = \mathbf{N}\cdot(\mathbf{C}\,\mathbf{N}) = \left[\mathbf{n}\cdot(\mathbf{B}^{-1}\mathbf{n})\right]^{-1} \tag{1.3.10}$$

where

$$\mathbf{C} = \mathbf{F}^{\mathrm{T}}\,\mathbf{F} \qquad \mathrm{C}_{\alpha\beta} = \mathrm{F}_{\alpha i}\,\mathrm{F}_{i\beta} \tag{1.3.11}$$

and

$$\mathbf{B} = \mathbf{F}\,\mathbf{F}^{T} \qquad B_{ij} = F_{i\alpha}\,F_{\alpha j} \tag{1.3.12}$$

are deformation tensors.
If **v** is the velocity,then the tensor **L** given by

$$\mathbf{L} = \frac{\partial \mathbf{v}}{\partial \mathbf{x}} = \operatorname{grad} \mathbf{v} \qquad L_{ij} = v_{i,j} \tag{1.3.13}$$

is called the tensor of velocity gradients.
L can be decomposed into a symmetric part written as **D** and an antisymmetric part written as **W**. Thus

$$\mathbf{L} = \mathbf{D} + \mathbf{W} \tag{1.3.14}$$

where

$$\mathbf{D} = \tfrac{1}{2}\left(\mathbf{L} + \mathbf{L}^{T}\right) \tag{1.3.15}$$

$$\mathbf{W} = \tfrac{1}{2}\left(\mathbf{L} - \mathbf{L}^{T}\right) \tag{1.3.16}$$

D is called the stretching tensor or rate of deformation tensor and **W** the spin tensor.

Let the element of area dS of a material surface in the reference configuration be carried into the element of surface ds in the current configuration.
The following relations hold

$$\mathbf{n}\,ds = J\,(\mathbf{F}^{-1})^{T}\,\mathbf{N}\,dS \tag{1.3.17}$$

From Eq. (1.3.17) we get

$$\left(\frac{ds}{dS}\right)^{2} = J^{2}\,\mathbf{N}\cdot\left(\mathbf{C}^{-1}\,\mathbf{N}\right) = J^{2}\left(\mathbf{n}\cdot(\mathbf{B}\,\mathbf{n})^{-1}\right) \tag{1.3.18}$$

and if we change in area occurs, then

$$\mathbf{N}\cdot\left(\mathbf{C}^{-1}\,\mathbf{N}\right) = J^{-2} \qquad \mathbf{n}\cdot\left(\mathbf{B}\,\mathbf{n}\right) = J^{2} \tag{1.3.19}$$

which imposes a restriction on **F**.

Consider the element of volume dV_r in the reference configuration which be carried into the element of volume dV.
We have

$$dV = \det \mathbf{F}\, dV_r \qquad dV = J\, dV_r \tag{1.3.20}$$

A motion in which $dV_r = dV$ i.e.

$$J = 1 \qquad (dV_r = dV) \tag{1.3.21}$$

is called an isochoric, or volume-preserving motion.
The condition for isochoric flow is

$$\text{tr}\,\mathbf{L} = \text{tr}\,\mathbf{D} = D_{ii} = \text{div}\,\mathbf{v} = v_{i,i} = 0 \tag{1.3.22}$$

1.4 Strain Tensors

The classical strain measures are the Almansi-Hamel strain **e** and the Green-Lagrange strain **E**.
The Almansi strain is defined by

$$2\mathbf{e} = \mathbf{I} - \mathbf{B}^{-1} \qquad 2e_{ij} = g_{ij} - X_{\alpha,i}\, X_{\alpha,j} \tag{1.4.1}$$

where g_{ij} is an important quantity characterizing the geometrical properties of space and is called metric tensor, $g_{ij} = \mathbf{g}_i \cdot \mathbf{g}_j$ where $(\cdot)$ is the scalar product of base vectors $\mathbf{g}_i$ and $\mathbf{g}_j$, $\mathbf{B}^{-1}$ is the inverse of **B** defined by Eq. (1.3.12) and the Green-Lagrange strain is defined by

$$2\mathbf{E} = \mathbf{C} - \mathbf{I} \qquad 2E_{\alpha\beta} = x_{k,\alpha}\, x_{k,\beta} - G_{\alpha\beta} \tag{1.4.2}$$

where $G_{\alpha\beta} = \mathbf{G}_\alpha \cdot \mathbf{G}_\beta$ is the scalar product of base vectors $\mathbf{G}_\alpha$ and $\mathbf{G}_\beta$ and **C** is defined by Eq. (1.3.11).
Consider the expression

$$2\,\mathbf{F}^T \mathbf{e}\, \mathbf{F} = \mathbf{F}^T \left[\mathbf{I} - \left(\mathbf{F}^{-1}\right)^T \mathbf{F}^{-1} \right] \mathbf{F} = \mathbf{C} - \mathbf{I} = 2\,\mathbf{E} \tag{1.4.3}$$

from which the relation between the two classical strain measures is given by

$$\mathbf{E} = \mathbf{F}^T\, \mathbf{e}\, \mathbf{F} \tag{1.4.4}$$

Noting that

$$\dot{\mathbf{F}} = \mathbf{L}\,\mathbf{F} \tag{1.4.5}$$

it follows that

$$\begin{aligned}\dot{\mathbf{E}} &= \tfrac{1}{2}\left(\dot{\mathbf{F}}^{\mathrm{T}}\,\mathbf{F} + \mathbf{F}^{\mathrm{T}}\,\dot{\mathbf{F}}\right) = \tfrac{1}{2}\left[\mathbf{F}^{\mathrm{T}}\left(\mathbf{F}^{\mathrm{T}}\right)^{-1}\dot{\mathbf{F}}^{\mathrm{T}}\,\mathbf{F} + \mathbf{F}^{\mathrm{T}}\,\dot{\mathbf{F}}\,\mathbf{F}^{-1}\,\mathbf{F}\right] \\ &= \tfrac{1}{2}\mathbf{F}^{\mathrm{T}}\left(\mathbf{L}^{\mathrm{T}} + \mathbf{L}\right)\mathbf{F} = \mathbf{F}^{\mathrm{T}}\,\mathbf{D}\,\,\mathbf{F}\end{aligned} \tag{1.4.6}$$

By (1.4.4) and (1.4.6) we get

$$\begin{aligned}\dot{\mathbf{E}} &= \mathbf{F}^{\mathrm{T}}\mathbf{D}\mathbf{F} = \dot{\mathbf{F}}^{\mathrm{T}}\mathbf{e}\,\mathbf{F} + \mathbf{F}^{\mathrm{T}}\mathbf{e}\,\mathbf{F} + \mathbf{F}^{\mathrm{T}}\mathbf{e}\,\dot{\mathbf{F}} \\ &= \mathbf{F}^{\mathrm{T}}\left(\dot{\mathbf{e}} + \mathbf{L}^{\mathrm{T}}\mathbf{e} + \mathbf{e}\,\mathbf{L}\right)\mathbf{F}\end{aligned} \tag{1.4.7}$$

Hence

$$\dot{\mathbf{e}} + \mathbf{L}^{\mathrm{T}}\,\mathbf{e} + \mathbf{e}\,\mathbf{L} = \mathbf{D} \tag{1.4.8}$$

Let a material point be displaced from the position **1** to the position **1'** and let the origins of the referential and spatial coordinates be at *0* and *o* respectively. The vectors **X**, **x**, **c** and displacement **u** are shown in the Figure 1-1. We have

$$\mathbf{x} + \mathbf{c} = \mathbf{X} + \mathbf{u} \tag{1.4.9}$$

At a neighboring point

$$(\mathbf{x} + \mathrm{d}\mathbf{x}) + \mathbf{c} = (\mathbf{X} + \mathrm{d}\mathbf{X}) + (\mathbf{u} + \mathrm{d}\mathbf{u}) \tag{1.4.10}$$

By (1.4.10)

$$\mathrm{d}\mathbf{x} = \mathrm{d}\mathbf{X} + \mathrm{d}\mathbf{u}$$

If

$$\mathrm{d}a^2 = \mathrm{d}\mathbf{x}\cdot\mathrm{d}\mathbf{x} \qquad \text{and} \qquad \mathrm{d}A^2 = \mathrm{d}\mathbf{X}\cdot\mathrm{d}\mathbf{X} \tag{1.4.11}$$

and, from Eq. (1.4.10)

$$da^2 - dA^2 = d\mathbf{x} \cdot d\mathbf{x} - d\mathbf{X} \cdot d\mathbf{X} = 2\, e_{ij}\, dx_i\, dx_j = 2\, d\mathbf{x} \cdot d\mathbf{u} - d\mathbf{u} \cdot d\mathbf{u} \qquad (1.4.12)$$

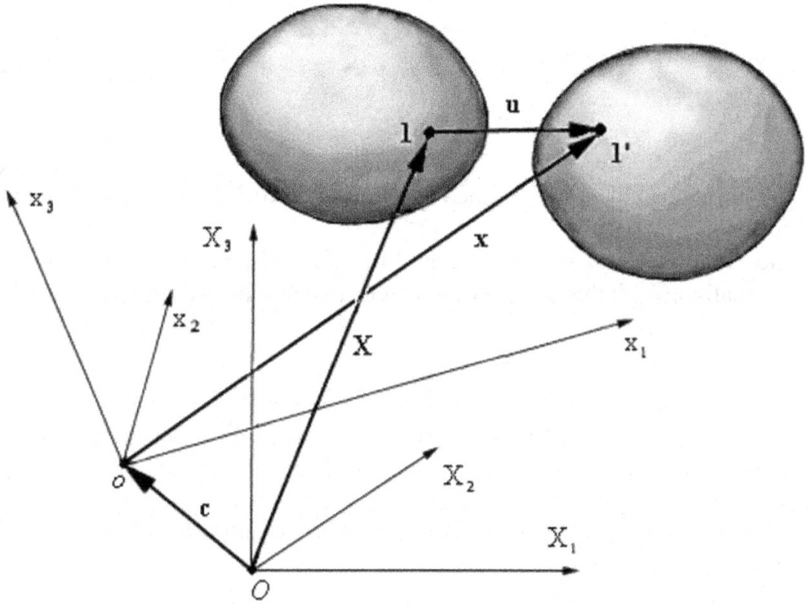

Figure 1-1. Displacement in reference and spatial configuration

We have

$$d\mathbf{u} = u_{i,j}\, g_i\, dx_j \qquad (1.4.13)$$

By (1.4.12) we get

$$e_{ij} = \tfrac{1}{2}\left(u_{j,i} + u_{i,j} - u_{k,i} u_{k,j}\right) \qquad (1.4.14)$$

Changing the roles of referential and spatial coordinates leads to

$$E_{\alpha\beta} = \tfrac{1}{2}\left(u_{\beta,\alpha} + u_{\alpha,\beta} + u_{\gamma,\alpha} u_{\gamma,\beta}\right) \qquad (1.4.15)$$

The strain tensors in Eqs. (1.4.14) and (1.4.15) are expressed in terms of the displacement gradients $u_{i,j}$ or $u_{\alpha,\beta}$.

If the displacement gradients are infinitesimal

$$\left\| u_{i,j} \right\| << 1 \tag{1.4.16}$$

then Eq. (1.4.14) reduces to

$$\varepsilon_{ij} = \tfrac{1}{2}\left(u_{j,i} + u_{i,j}\right) \tag{1.4.17}$$

The infinitesimal strain defined by Eq. (1.4.16) is known as Cauchy strain tensor.

The relation (1.4.16) is called the hypothesis of infinitesimal transformation. With this assumption the Green-Lagrange strain is

$$E_{\alpha\beta} \cong g_{i\alpha} g_{j\beta} \varepsilon_{ij} \tag{1.4.18}$$

and the principal values of both strains are the same. Thus for infinitesimal transformation there is no difference between spatial and referential strain measures.

Consider the elements **dX** which defined a sphere of radius R_o at **X**.
If radius is constant, then

$$dA^2 = G_{\alpha\beta}\, dX_\alpha\, dX_\beta = R_o^2 \tag{1.4.19}$$

Since

$$da^2 = d\mathbf{x} \cdot d\mathbf{x} \qquad\qquad dA^2 = d\mathbf{X} \cdot d\mathbf{X} \tag{1.4.20}$$

Eq. (1.3.10) can be rearranged to the form

$$dA^2 = \left(B^{-1}\right)_{ij} dx_i dx_j \tag{1.4.21}$$

Substituting dA from Eq. (1.4.19) we get

$$R_o^2 = G_{\alpha\beta} dX_\alpha dX_\beta = \left(B^{-1}\right)_{ij} dx_i dx_j \tag{1.4.22}$$

The rotation is characterized by the **R** tensor

$$R_{ij} = \tfrac{1}{2}\left(u_{i,j} - u_{j,i}\right) \tag{1.4.23}$$

The dual or axial vector is given by

$$\begin{aligned}\omega^i &= \tfrac{1}{2}\varepsilon_{ijk}R_{jk} = \tfrac{1}{4}\varepsilon_{ijk}\left(u_{j,k} - u_{k,j}\right) = \tfrac{1}{4}\varepsilon_{ijk}u_{j,k} - \tfrac{1}{4}\varepsilon_{ikj}u_{j,k} \\ &= \tfrac{1}{2}\varepsilon_{ijk}u_{j,k} = \tfrac{1}{2}(\mathrm{curl}\,u)_i\end{aligned} \tag{1.4.24}$$

The symbol ε_{ijk} is given by

$$\varepsilon_{ijk} = e_{ijk}\sqrt{g} \tag{1.4.25}$$

where $g = \det g_{ij}$ and e_{ijk} is so-called a permutation symbol defined by the set of equations

$$\begin{aligned} e_{123} &= e_{231} = e_{312} = +1 \\ e_{132} &= e_{321} = e_{213} = -1 \end{aligned} \tag{1.4.26}$$

and all other values of i, j and k make e_{ijk} zero.
By (1.4.17) and (1.4.23)

$$u_{i,j} = \varepsilon_{ij} + R_{ij} \tag{1.4.27}$$

$$u_{i,j} = \varepsilon_{ij} + \varepsilon_{ijk}\omega^k \tag{1.4.28}$$

If the space is Euclidean, making use of Eq. (1.4.17)

$$\begin{aligned} \varepsilon_{ij,kl} &= \tfrac{1}{2}\left(u_{j,ikl} + u_{i,jkl}\right) \\ \varepsilon_{kl,ij} &= \tfrac{1}{2}\left(u_{l,kij} + u_{k,lij}\right) \\ \varepsilon_{ik,jl} &= \tfrac{1}{2}\left(u_{k,ijl} + u_{i,kjl}\right) \\ \varepsilon_{jl,ik} &= \tfrac{1}{2}\left(u_{l,jik} + u_{j,lik}\right) \end{aligned} \tag{1.4.29}$$

By Eqs. (1.4.29) we find the so-called compatibility conditions

$$\varepsilon_{ij,kl} + \varepsilon_{kl,ij} - \varepsilon_{ik,jl} - \varepsilon_{jl,ik} = 0 \tag{1.4.30}$$

This set of equations reduces to six independent equations when the symmetry of $\boldsymbol{\varepsilon}$ is taken into account.

1.5 Particulate and Material Derivatives

Let G (**X**, t) be a field in a Lagrange description. The time derivative of a field G (**X**, t) multiplied by the infinitesimal time interval dt is equal to the variation between times t and t + dt of the function G (**X**, t), which would be recorded by an observer attached to the material particle which is located in the reference configuration by position vector **x**. In terms of Euler variables the particulate derivative with respect to the material is the total time derivative of the field g [**x** (**X**, t)] = G (**X**, t)

$$\frac{dg}{dt} = \frac{\partial g}{\partial t} + \operatorname{grad} g \cdot \mathbf{v} \tag{1.5.1}$$

For example, substituting **v** for g in Eq. (1.5.1) the expression for acceleration ***a*** in Euler variables can be written as

$$\boldsymbol{a} = \frac{d\,\mathbf{v}}{dt} = \frac{\partial\,\mathbf{v}}{\partial t} + \operatorname{grad} \mathbf{v} \cdot \mathbf{v} \tag{1.5.2}$$

Consider the volume integral

$$\mathcal{G} = \int_V g(\mathbf{x}, t)\, dV \tag{1.5.3}$$

where g (**x**, t) is the volume density in the current configuration.

Let G(**X**, t) be the volume Lagrangian density, then the corresponding Eulerian density is g(**x**, t)

$$g(\mathbf{x}, t)\, dV = G(\mathbf{X}, t)\, dV_r \tag{1.5.4}$$

where the volume V_r refers to the reference configuration of the volume V in the current configuration.

The time derivation of Eq. (1.5.3) throughout the relation (1.5.4) has the form

$$\frac{d\mathcal{G}}{dt} = \int_{V_r} \frac{dG}{dt}\, dV_r \tag{1.5.5}$$

The expression (1.5.5) represents the particulate derivative of the volume integral $\mathcal{G}$ in terms of Lagrange variables.

In order to obtain the particulate time derivative in Eulerian variables the equality (1.5.4) must be taken.
By Eqs. (1.3.20) and (1.5.1) and the expression

$$\operatorname{div}(g \otimes \mathbf{v}) = g \operatorname{div} \mathbf{v} + \operatorname{grad} g \cdot \mathbf{v} \tag{1.5.6}$$

we get

$$\frac{dG}{dt} dV_r = \left[\frac{\partial g}{\partial t} + \operatorname{div}(g \otimes \mathbf{v}) \right] dV \tag{1.5.7}$$

The particulate derivative of the volume integral $\mathcal{G}$ with respect to the material derived in Eulerian variables yields

$$\frac{d\mathcal{G}}{dt} = \int_V \left[\frac{\partial g}{\partial t} + \operatorname{div}(g \otimes \mathbf{v}) \right] dV \tag{1.5.8}$$

The alternative form of Eq. (1.5.8) can be obtained by using the divergence theorem. Thus

$$\frac{d\mathcal{G}}{dt} = \int_V \frac{\partial g}{\partial t} dV_r + \int_a g\, \mathbf{v} \cdot \mathbf{n}\, da \tag{1.5.9}$$

where a is the surface of the volume V.

The first term of the right-hand side of Eq. (1.5.9) represents the variation of the field g between time t and t + dt in volume V, and the other one due to the movement of the same material volume.

The material derivative is used to determine the variation between time t and t + dt of any physical quantity attached to the whole material, which is contained at time t in volume V. The particulate derivative with respect to the material only partially takes into account this variation. It ignores any mass particles leaving the volume V, which is followed in the material movement.

In the infinitesimal time interval dt = Dt the variation of quantity g attached to the whole matter at time t in the volume V as the variation $D\mathcal{G}$ of the integral $\mathcal{G}$ given by Eq. (1.5.5) involves

$$\frac{D\mathcal{G}}{Dt} = \frac{d}{dt} \int_V g(\mathbf{x}, t)\, dV \tag{1.5.10}$$

Substituting Eq. (1.5.9) of the particulate derivatives with respect to the material of the volume integral we obtain

$$\frac{D\mathcal{G}}{Dt} = \int_V \frac{\partial g}{\partial t} dV + \int_a g\, \mathbf{v} \cdot \mathbf{n}\, da \qquad (1.5.11)$$

1.6 Mass Conservation

Let ρ be a scalar field defined on body B, where ρ is referred to the mass density of the material of which the body is composed. The mass contained in the infinitesimal volume dV is equal to ρdV. We assume no overall mass creation, which implies the global mass balance

$$\frac{D}{Dt} \int_V \rho dV = 0 \qquad (1.6.1)$$

By Eqs. (1.5.11) and (1.6.1) the mass balance reads

$$\int_V \frac{\partial \rho}{dt} dV + \int_a \rho\, \mathbf{v} \cdot \mathbf{n}\, da = 0 \qquad (1.6.2)$$

Applying the divergence theorem to Eq. (1.6.2) we get the local mass balance equation or continuity equation

$$\frac{\partial \rho}{\partial t} + \operatorname{div}(\rho\, \mathbf{v}) = 0 \qquad (1.6.3)$$

or equivalently

$$\frac{\partial \rho}{\partial t} + \rho \operatorname{div} \mathbf{v} = 0 \qquad (1.6.4)$$

The mass ρdV, which is contained in volume dV, may be written as

$$\rho\, dV = \rho_o\, dV_r \qquad (1.6.5)$$

where ρ_o and ρ are the mass densities in the reference and current configurations respectively.

The mass conservation may now be written as

$$\frac{D}{Dt}\int_{V_r}\rho_o dV_r = 0 \qquad (1.6.6)$$

From the transport formula it is evident that the conservation of mass in Lagrange variables can be expressed in the form

$$J\rho = \rho_o \qquad (1.6.7)$$

Chapter 2

STRESS TENSOR

2.1 Momentum

The linear momentum $\boldsymbol{p}$ of the material occupying the volume V in the configuration at time t by a moving material body B_t is defined by the relation

$$\boldsymbol{p} = \int_V \rho \mathbf{v} \, dV \tag{2.1.1}$$

where **v** is the velocity of the material particle at the point **x**.

Another important quantity is the angular momentum $\boldsymbol{H}$ which is defined by the relation

$$\boldsymbol{H} = \int_V (\mathbf{x} \times \mathbf{v}) \rho \, dV \qquad H_i = \int_V \varepsilon_{ijk} x_j v_k \rho \, dV \tag{2.1.2}$$

where **x** is the position of a representative point of V relative to an origin *o* and $\boldsymbol{H}$ being the angular momentum with respect to *o*.
Equations of motion in Euler description is given in the form of the two principles. The first one is: the rate of change of linear momentum $\boldsymbol{p}$ is equal to the total applied force $\boldsymbol{F}$

$$\dot{\boldsymbol{p}} = \boldsymbol{F} \tag{2.1.3}$$

The second one says: the rate of change of angular momentum $\boldsymbol{H}$ is equal to the total applied torque $\boldsymbol{\Gamma}$, i.e.

$$\boldsymbol{H} = \boldsymbol{\Gamma} \tag{2.1.4}$$

Two types of external forces are assumed to act on a body B. The first are body forces, such as gravity acting on material element throughout the body and can be described by a vector field **f** which is referred to as the body force per unit mass and which is defined on the configurations of B. The second are surface forces such as friction which act on the surface elements of area and can be described by a vector **t** which is referred to as the surface fraction per unit area and which is defined on the surface a of the material.

The total force $\boldsymbol{F}$ is defined as

$$\boldsymbol{F} = \int_V \rho \mathbf{f}\, dV + \int_a \mathbf{t}\, da \tag{2.1.5}$$

The second integral on the right-hand side of Eq. (2.1.5) represents the contribution to $\boldsymbol{F}$ of the contact force acting on the boundary a of the arbitrary material region of the volume V.

The total torque $\boldsymbol{\Gamma}$ about the origin o is defined as

$$\boldsymbol{\Gamma} = \int_V \rho(\mathbf{x} \times \mathbf{f})\, dV + \int_a (\mathbf{x} \times \mathbf{t})\, da \tag{2.1.6}$$

The second integral on the right-hand side of Eq. (2.1.6) represents the contribution to $\boldsymbol{\Gamma}$ of the contact forces acting on the boundary a of the arbitrary region.

2.2 The Stress Vector

In the course of deformation of solids, on account of volume and geometrical shape changes, interactions between molecules come into being that oppose these changes. The concept that the action of the rest of the material upon any volume element of it was of the same form as distributed surface forces was introduced by Cauchy. In other words the effect of the material on one side of a surface a, on the material on the other side is equivalent to a distribution of force vector **t** per unit area. The force vector per unit area **t** is also called the surface traction or stress vector.

The stress vector **t** at any point is associated with an element of surface **da** having a unit normal **n**.

The balance of linear momentum for the body shown in Fig. 2.1. is

$$\mathbf{t}(\mathbf{n})\, da + \mathbf{t}(-\mathbf{n})\, da = d\, da\, \rho\, \ddot{\mathbf{x}} \tag{2.2.1}$$

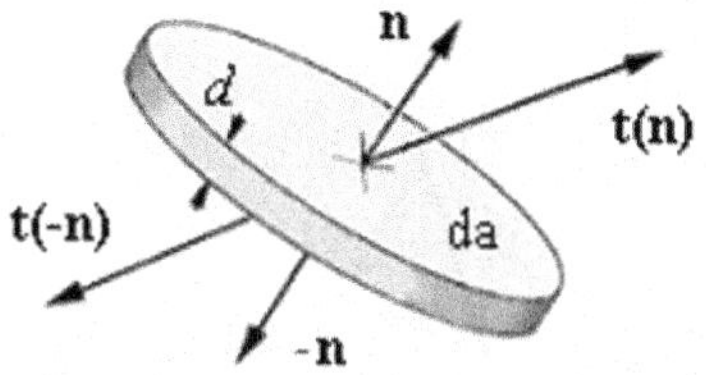

Figure 2-1. The stress vector on body surface

when the dependence of the stress vector on direction is written in the form $\mathbf{t}(\mathbf{n})$ and $\mathrm{da}\rho\,\ddot{\mathbf{x}}$ are peripheral forces.

In the case when *d* tends to zero, then by Eq. (2.2.1)

$$\mathbf{t}(\mathbf{n}) = -\mathbf{t}(-\mathbf{n}) \tag{2.2.2}$$

The above has the same form as Newton's third law, i.e. action and reaction are equal and opposite.

Consider a tetrahedron shown in Fig. 2.2. The sides of the tetrahedron are taken along the coordinate lines through the point **1** with the vectorial elements of length being given by $\mathbf{12} = \mathbf{g}_1 dx_1$, $\mathbf{13} = \mathbf{g}_2 dx_2$, $\mathbf{14} = \mathbf{g}_3 dx_3$, where $\mathbf{g}_i$ are the base vectors.

Denote the unit outward normals to the faces **134**, **142**, **123** and **234** as $-\mathbf{n}_1$, $-\mathbf{n}_2$, $-\mathbf{n}_3$ and $\mathbf{n}$, the tractions on the faces be $-\mathbf{t}_1$, $-\mathbf{t}_2$, $-\mathbf{t}_3$ and $\mathbf{t}$, and the areas of the faces be da_1, da_2, da_3 and da respectively.

We have

$$\mathbf{13} \times \mathbf{14} = \mathbf{n}_1\,\mathrm{da}_1 \qquad \mathbf{14} \times \mathbf{12} = \mathbf{n}_2\,\mathrm{da}_2$$

$$\mathbf{12} \times \mathbf{13} = \mathbf{n}_3\,\mathrm{da}_3 \qquad \mathbf{23} \times \mathbf{24} = \mathbf{n}\,\mathrm{da} \tag{2.2.3}$$

We get

$$\mathbf{n}\,\mathrm{da} = (\mathbf{13} - \mathbf{12}) \times (\mathbf{14} - \mathbf{12}) = \mathbf{n}_1\,\mathrm{da}_1 + \mathbf{n}_2\,\mathrm{da}_2 + \mathbf{n}_3\,\mathrm{da}_3$$

The unit vectors are given by

$$\mathbf{n}_1 = \frac{\mathbf{g}_1}{\sqrt{g_{11}}} \qquad \mathbf{n}_2 = \frac{\mathbf{g}_2}{\sqrt{g_{22}}} \qquad \mathbf{n}_3 = \frac{\mathbf{g}_3}{\sqrt{g_{33}}} \tag{2.2.4}$$

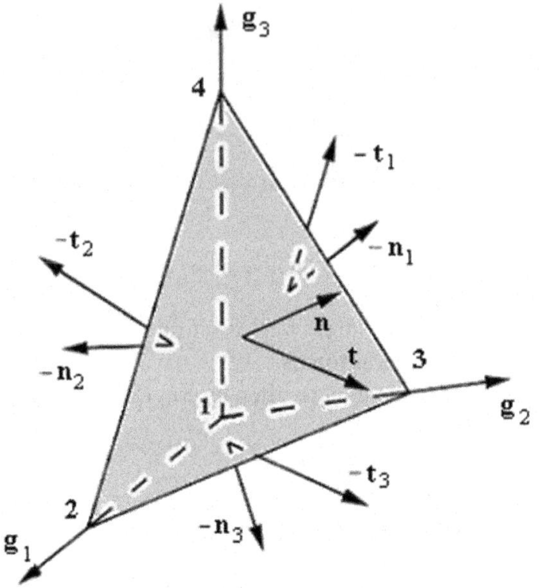

Figure 2-2. Forces acting on tetrahedron body

Substituting Eq. (2.2.4) into Eq. (2.2.3) we have

$$\mathbf{n}\,\mathrm{da} = \sum_{i=1}^{3}\left(\frac{\mathrm{da}_i}{\sqrt{g_{ii}}}\right)\mathbf{g}_i \tag{2.2.5}$$

But

$$\mathbf{n} = n_i\,\mathbf{g}_i \tag{2.2.6}$$

and we get

$$\mathrm{da}_i = n_i\sqrt{g_{ii}}\,\mathrm{da} \tag{2.2.7}$$

The rate of change of linear momentum for the tetrahedron is

$$\mathbf{t}\,\mathrm{da} - \mathbf{t}_i\,\mathrm{da}_i = \rho\,\dot{\mathbf{v}}\,d\mathrm{da} \tag{2.2.8}$$

where **v** is the velocity of the tetrahedron and *d* is the perpendicular distance from **1** to the surface **234**.

By Eqs. (2.2.8) and (2.2.7) we get

$$\mathbf{t} = \sum_{i=1}^{3} \mathbf{t}_i \, n_i \sqrt{g_{ii}} \tag{2.2.9}$$

Since

$$\mathbf{t}_i \sqrt{g_{ii}} = T_{ij} \, \mathbf{g}_j \qquad \mathbf{t} = t_j \, \mathbf{g}_j \tag{2.2.10}$$

it is evident, that Eq. (2.2.9) implies that **t** must be a linear transformation of **n**; i.e.

$$\mathbf{t} = \mathbf{T}^T \cdot \mathbf{n} \tag{2.2.11}$$

which relates the stress vector to the unit normal **n**, it being noted that **T** is defined on the configurations of the material body and in particular does not depend upon **n**.

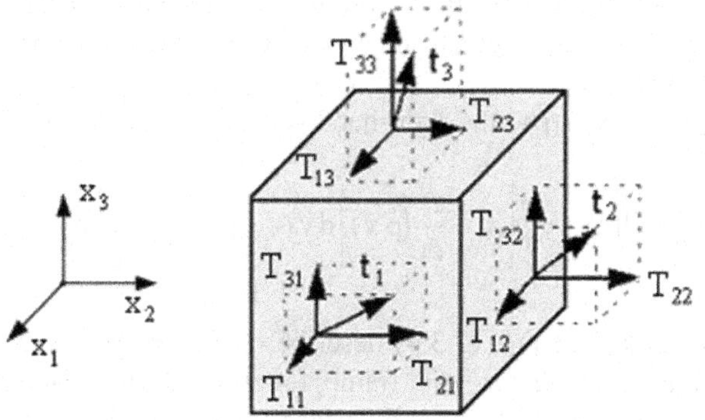

Figure 2-3. The physical meaning of the Cauchy stress tensor in the Euclidean space

The tensor **T** is called Cauchy's stress tensor or simply the stress tensor.

2.3 Momentum Balance

Eq. (2.1.5) can be expressed by way of Eqs. (2.1.1) and (2.1.2) in the form

$$\frac{D}{Dt}\int_V \rho\,\mathbf{v}\,dV = \int_V \rho\,\mathbf{f}\,dV + \int_a \mathbf{t}\,da \tag{2.3.1}$$

$$\frac{D}{Dt}\int_V [\mathbf{x}\times\rho\,\mathbf{v}]dV = \int_V \mathbf{x}\times\rho\,\mathbf{f}\,dV + \int_a \mathbf{x}\times\mathbf{t}\,da \tag{2.3.2}$$

Equation (2.3.1) shows that the creation rate due to the external forces acting on material is the instantaneous time derivatives of the linear momentum of the material in the volume V.

Equation (2.3.2) states the same, but concerns the angular momentum. In Eulerian expression of the material derivative of the volume integral, the left-hand side of Eq. (2.3.1) can be expressed as

$$\frac{D}{Dt}\int_V \rho\,\mathbf{v}\,dV = \int_V \frac{\partial}{\partial t}(\rho\,\mathbf{v})dV + \int_a [\rho\,\mathbf{v}\,\mathbf{v}\cdot\mathbf{n}]da \tag{2.3.3}$$

In Eq. (2.3.3) the volume integral corresponds to the time variation of linear momentum in elementary volume V, and the surface integral corresponds to the momentum carried away by the material leaving the same geometrical volume.

Similarly, if $g = \mathbf{x} \times (\rho\mathbf{v})$ the right-hand side of (2.3.2) can be rewritten as

$$\frac{D}{Dt}\int_V [\mathbf{x}\times(\rho\,\mathbf{v})]dV = \int_V \left[\mathbf{x}\times\frac{\partial}{\partial t}(\rho\,\mathbf{v})\right]dV + \int_a [(\mathbf{x}\times\rho\,\mathbf{v})\mathbf{v}\cdot\mathbf{n}]da \tag{2.3.4}$$

Altogether Eqs. (2.3.1) – (2.3.4) describe the Euler theorem, which can be stated as follows. In a referential frame, for any material subdomain V the resultant of the elementary body and surface forces and the resultant of the corresponding elementary moments are respectively equal to the resultant and the overall moment of vectors

$$\frac{\partial}{\partial t}(\rho\,\mathbf{v})dV \tag{2.3.5}$$

distributed in the volume V, and the resultant and the overall moment of vectors

$$[(\rho\,\mathbf{v})\mathbf{v}\cdot\mathbf{n}]da \tag{2.3.6}$$

distributed on the surface a of the domain V.
By the divergence theorem, Eqs. (2.3.3) and (2.3.4) are rewritten in the form

$$\frac{D}{Dt}\int_V \rho\,\mathbf{v}\,dV = \int_V \frac{\partial}{\partial t}(\rho\,\mathbf{v})dV + \int_V \mathrm{div}[\rho\,\mathbf{v}\otimes\mathbf{v}]dV \tag{2.3.7}$$

$$\frac{D}{Dt}\int_V \mathbf{x}\times\rho\,\mathbf{v}\,dV = \int_V \mathbf{x}\times\frac{\partial}{\partial t}(\rho\,\mathbf{v})dV + \int_V \mathbf{x}\times\mathrm{div}[\rho\,\mathbf{v}\otimes\mathbf{v}]dV \tag{2.3.8}$$

By the relation

$$\mathrm{div}(\mathbf{v}^*\otimes\mathbf{v}) = \mathrm{grad}\,\mathbf{v}^*\cdot\mathbf{v} + \mathbf{v}^*\,\mathrm{div}\,\mathbf{v} \tag{2.3.9}$$

we get

$$\frac{\partial}{\partial t}(\rho\,\mathbf{v}) + \mathrm{div}[\rho\,\mathbf{v}\otimes\mathbf{v}] = \rho\left[\frac{\partial\,\mathbf{v}}{\partial t} + \mathrm{grad}\,\mathbf{v}\cdot\mathbf{v}\right] + \mathbf{v}\left[\frac{\partial\rho}{\partial t} + \mathrm{div}(\rho\,\mathbf{v})\right] \tag{2.3.10}$$

By Eq. (1.5.2) we get

$$\frac{\partial}{\partial t}(\rho\,\mathbf{v}) + \mathrm{div}[\rho\,\mathbf{v}\otimes\mathbf{v}] = \rho\frac{d\,\mathbf{v}}{dt} \tag{2.3.11}$$

The expressions (2.3.7), (2.3.8) and (2.3.11) give a new form of the momentum balance namely the dynamic theorem

$$\begin{gathered}\int_V \rho\frac{d\,\mathbf{v}}{dt}dV = \int_V \rho\,\mathbf{f}\,dV + \int_a \mathbf{t}\,da \\ \int_V \mathbf{x}\times\rho\frac{d\,\mathbf{v}}{dt}dV = \int_V \mathbf{x}\times\rho\,\mathbf{f}\,dV + \int_a \mathbf{x}\times\mathbf{t}\,da\end{gathered} \tag{2.3.12}$$

which states that the resultant of the elementary body and surface forces and the resultant of the corresponding elementary moments are equal respectively to the resultant and the overall movement of the elementary dynamic forces.
Substituting from Eq. (2.2.11) for **t** in Eq. (2.3.1) and rearranging by way of Eq. (1.6.4) gives

$$\int_V \rho\left(\frac{d\mathbf{v}}{dt} - \mathbf{f}\right)dV = \int_a \mathbf{Tn}\, da \tag{2.3.13}$$

Using the divergence theorem Eq. (2.3.13) can be rearranged into the form

$$\int_V \left[\rho\left(\frac{d\mathbf{v}}{dt} - \mathbf{f}\right) - \operatorname{div}\mathbf{T}\right] dV = 0 \tag{2.3.14}$$

The condition that Eq. (2.3.14) holds for all arbitrary material regions V leads to the field equation form of the balance of linear momentum

$$\rho\frac{d\mathbf{v}}{dt} = \operatorname{div}\mathbf{T} + \rho\mathbf{f} \tag{2.3.15}$$

which is Cauchy's equation of motion.

2.4 Properties of the Stress Tensor

Consider the components of the stress tensor T_{ij} at some point $\mathbf{x}$ in the current configuration B_t. By Eqs. (2.2.9) and (2.2.11) it follows that

$$T_{ij} = \mathbf{g}_i \cdot \left(\mathbf{T}\mathbf{g}_j\right) = \mathbf{g}_j \cdot \left(\mathbf{T}^T \mathbf{g}_i\right) = \sqrt{g_{ii}}\, \mathbf{t}_i \cdot \mathbf{g}_j \tag{2.4.1}$$

From Eq. (2.4.1) it is seen that T_{ij} can be interpreted as the j component of the force per unit area in B_t acting on a surface segment which outward normal at $\mathbf{x}$ is in the i direction. At a given point $\mathbf{x}$, there exists in general three mutually perpendicular principal directions $\mathbf{n}_a\,(a = 1,2,3)$, with components $(n_a)_i\,(i = 1,2,3)$ which are the solutions of equation

$$\left(T_{ij} - \sigma\delta_{ij}\right)n_j = 0 \tag{2.4.2}$$

corresponding to the roots $\sigma_a\,(a = 1,2,3)$ which are called the principal stresses.

The stress tensor at $\mathbf{x}$ have the representations

$$\mathbf{T} = \sum_{a=1}^{3} \sigma_a\, \mathbf{n}_a \otimes \mathbf{n}_a \tag{2.4.3}$$

The stress vector can be resolved into a component normal to the surface element $\mathbf{t}_{(n)}$ and a component $\mathbf{t}_{(s)}$ tangential to the surface element. The component $\mathbf{t}_{(n)}$ is called the normal stress and $\mathbf{t}_{(s)}$ the shearing stress.
The magnitude $\sigma_{(n)}$ of $\mathbf{t}_{(n)}$ is given by

$$\sigma_{(n)} = \left|\mathbf{t}_{(n)}\right| = \mathbf{t} \cdot \mathbf{n} = t_i n_i = T_{ij} n_i n_j \tag{2.4.4}$$

Using Eq. (2.4.4), $\sigma_{(n)}$ can be written in the form

$$\sigma_{(n)} = \sigma_1 n_1^2 + \sigma_2 n_2^2 + \sigma_3 n_3^2 \tag{2.4.5}$$

The shearing stress $\mathbf{t}_{(s)}$ can be written as

$$\mathbf{t}_{(s)} = \mathbf{t} - \mathbf{t}_{(n)} \tag{2.4.6}$$

and its magnitude $\tau_{(s)}$ is given by

$$\begin{aligned} \tau_{(s)}^2 &= \mathbf{t}_{(s)} \cdot \mathbf{t}_{(s)} = t^2 - \sigma_{(n)}^2 \\ &= \sigma_1^2 n_1^2 + \sigma_2^2 n_2^2 + \sigma_3^2 n_3^2 - \left(\sigma_1 n_1^2 + \sigma_2 n_2^2 + \sigma_3 n_3^2\right)^2 \end{aligned} \tag{2.4.7}$$

Let $n_1 = 0$ and $n_2^2 = n_3^2 = \frac{1}{2}$. Substituting in Eq. (2.4.7) we get

$$\tau_{(s)} = \frac{1}{2}\left(\sigma_2 - \sigma_3\right) \tag{2.4.8}$$

The values of $\tau_{(s)}$ satisfying such relations are called the principal shearing stresses. The maximum shearing stress is thus equal to half the difference between the maximum and minimum principal stresses and acts on a plane which makes an angle of 45° with the directions of these principal stresses.
The characteristic equation for $\mathbf{T}$ is

$$\sigma_a^3 - \sigma_a^2 J_{1T} + \sigma_a J_{2T} - J_{3T} = 0 \qquad \text{for } a = 1, 2, 3 \tag{2.4.9}$$

when J_{1T}, J_{2T}, J_{3T} are called the invariants.
Taking the coordinate axes as the principal directions it can be shown that $\mathbf{T}$ has the form

$$\left[T_{ij}\right]=\begin{bmatrix}\sigma_1 & 0 & 0\\ 0 & \sigma_2 & 0\\ 0 & 0 & \sigma_3\end{bmatrix} \tag{2.4.10}$$

The invariants can be written as

$$\begin{aligned} \mathbf{J}_{1\mathbf{T}} &= \sigma_1+\sigma_2+\sigma_3\\ \mathbf{J}_{2\mathbf{T}} &= \sigma_1\sigma_2+\sigma_2\sigma_3+\sigma_3\sigma_1\\ \mathbf{J}_{3\mathbf{T}} &= \sigma_1\sigma_2\sigma_3 \end{aligned} \tag{2.4.11}$$

The deviator of the stress tensor $\mathbf{T}'$ is defined as

$$\mathbf{T}'=\mathbf{T}-\frac{1}{3}(\mathrm{tr}\,\mathbf{T})\mathbf{I} \tag{2.4.12}$$

The expression $T^0=\frac{1}{3}\mathrm{tr}\,\mathbf{T}$ is called the mean normal stress.

2.5 The Virtual Work Rate

Let $\mathbf{v}^*$ be any velocity field. Owing to its definition, the strain work rate R_{SR} $(\mathbf{v}^*)$ associated with any velocity field $\mathbf{v}^*$ and relative to material domain V, reads

$$R_{SR}(\mathbf{v}^*)=\int_V \mathbf{v}^*\cdot\rho\mathbf{f}\,dV+\int_a \mathbf{v}^*\cdot\mathbf{t}\,da-\int_V \mathbf{v}^*\cdot\rho\boldsymbol{a}dV \tag{2.5.1}$$

The last term of Eq. (2.5.1) represents the work rate of inertia forces and the first two integrals of the right-hand side of Eq. (2.5.1) represent the work rate of the external body and surface forces.

The symmetry of the stress tensor $\mathbf{T}$ and the divergence theorem applied yield the identity

$$\int_a \mathbf{v}^*\cdot\mathbf{T}\cdot\mathbf{n}\,da=\int_V(\mathbf{T}\,\mathbf{D}^*+\mathbf{v}^*\cdot\mathrm{div}\,\mathbf{T})dV \tag{2.5.2}$$

where $\mathbf{T}\mathbf{D}^*=T_{ij}D^*_{ij}$ and $\mathbf{D}^*$ is the strain rate associated with velocity field $\mathbf{v}^*$

$$2\mathbf{D}^* = \mathbf{L}^* + (\mathbf{L}^*)^T \tag{2.5.3}$$

where $\mathbf{L}^*$ = grad $\mathbf{v}^*$.

By the motion equation (2.3.15) and Eq. (2.5.2) the strain work rate $R_{SR}(\mathbf{v}^*)$ defined by (2.5.1) can be written as

$$R_{SR}(\mathbf{v}^*) = \int_V dr_{SR}(\mathbf{v}^*) \qquad dr_{SR}(\mathbf{v}^*) = \mathbf{T}\,\mathbf{D}^*\,dV \tag{2.5.4}$$

where dr_{SR} $(\mathbf{v}^*)$ is the infinitesimal strain work rate of the elementary material domain dV.

If velocity field $\mathbf{v}^*$ is equal to velocity $\mathbf{v}$ i.e. it is an actual velocity of the material particle, then

$$dr_{SR}(\mathbf{v}) = \mathbf{T}\,\mathbf{D}\,dV \tag{2.5.5}$$

The work rate of the internal forces denoted as R_{IF} is the opposite of the strain work rate R_{SR} i.e.

$$R_{IF}(\mathbf{v}^*) = -R_{SR}(\mathbf{v}^*) \qquad dr_{IF}(\mathbf{v}^*) = -dr_{SR} \tag{2.5.6}$$

The strain work rate of external forces is defined as

$$R_{EF}(\mathbf{v}^*) = \int_V \mathbf{v}^* \cdot \rho\mathbf{f}\,dV + \int_a \mathbf{v}^* \cdot \mathbf{t}\,da \tag{2.5.7}$$

and the work rate of inertia forces is

$$R_{IN}(\mathbf{v}^*) = -\int_V \mathbf{v}^* \cdot \rho\boldsymbol{a}\,dV \tag{2.5.8}$$

The virtual work rate theorem is simply rewriting Eq. (2.5.1) in terms of (2.5.7.) and (2.5.8)

$$R_{EF}(\mathbf{v}^*) + R_{IN}(\mathbf{v}^*) + R_{IF}(\mathbf{v}^*) = 0 \tag{2.5.9}$$

for every volume V and velocity $\mathbf{v}^*$.

It states that for any actual or virtual velocity field $\mathbf{v}^*$ and for any material domain V, the sum of the external forces R_{EF} $(\mathbf{v}^*)$, inertia forces R_{IN} $(\mathbf{v}^*)$ and internal forces R_{IF} $(\mathbf{v}^*)$ is equal to zero.

2.6 The Piola-Kirchhoff Stress Tensor

The strain work rate dr_{SR} defined by (2.5.5) is independent of the choice of the coordinate system used to describe the motion. The definition (2.5.5) corresponds to a Eulerian description.
By Eqs. (1.3.20) and (1.3.22) the expression (2.5.5) can be written as

$$dr_{SR}(\mathbf{v}) = \mathbf{T}\,\mathbf{D}\,dV = J\left(\mathbf{F}^{-1}\cdot\mathbf{T}\cdot\left(\mathbf{F}^{-1}\right)^{T}\right)\frac{d\,\mathbf{E}}{dt}dV_r \tag{2.6.1}$$

where $d\mathbf{E}/dt$ is the transformation of $\mathbf{D}$ in the reference configuration

$$\mathbf{D} = \left(\mathbf{F}^{-1}\right)^{T}\cdot\frac{d\,\mathbf{E}}{dt}\cdot\mathbf{F}^{-1} \tag{2.6.2}$$

Equation (2.6.1) serves to introduce the symmetric Piola-Kirchhoff stress tensor $\mathbf{S}$ defined by

$$\mathbf{S} = J\left(\mathbf{F}^{-1}\cdot\mathbf{T}\cdot\left(\mathbf{F}^{-1}\right)^{T}\right) \tag{2.6.3}$$

The tensors $\mathbf{D}$ and $\frac{d\,\mathbf{E}}{dt}$ represent the same material tensor in different configurations.

Using Eqs. (2.6.1) and (2.6.3), the Lagrangian description of the elementary and overall strain work rates $dr_{SR}(\mathbf{v})$ and $R_{SR}(\mathbf{v})$ are given, respectively, by

$$dr_{SR}(\mathbf{v}) = \mathbf{S}\frac{d\,\mathbf{E}}{dt}dV_r \qquad R_{SR}(\mathbf{v}) = \int_{V_r} dr_{SR}(\mathbf{v})dV_r \tag{2.6.4}$$

The definition (2.6.3) of Piola-Kirchhoff stress tensor $\mathbf{S}$ gives

$$\mathbf{F}\cdot\mathbf{S}\cdot\mathbf{N}\,dA = \mathbf{T}\cdot\mathbf{n}\,da \tag{2.6.5}$$

The dynamic theorem Eq. (2.3.12) throughout Eqs. (2.2.11) and (2.6.5) expresses the equality between the dynamical resultant of external forces for the material domain V in a Lagrangian description as

$$\int_A \mathbf{F}\cdot\mathbf{S}\cdot\mathbf{N}\,dA + \int_{V_r}\rho_o\left(\mathbf{f} - \frac{d\,\mathbf{v}}{dt}\right)dV_r = 0 \tag{2.6.6}$$

where the domain V_r and the surface A enclosing this domain in reference configuration correspond to the domain V and the surface a enclosing it in the current configuration.

The equation of motion in a Lagrangian description is obtained by the transformation of the surface integral into volume integral throughout the divergence theorem

$$\mathrm{Div}(\mathbf{F}\cdot\mathbf{S})+\rho_o\left(\mathbf{f}-\frac{d\,\mathbf{v}}{dt}\right)=0 \tag{2.6.7}$$

$$\frac{\partial}{\partial X_\alpha}\left(\frac{\partial x_i}{\partial X_\beta}S_{\beta\alpha}\right)+\rho_o\left(f_i-\frac{dv_i}{dt}\right)=0 \tag{2.6.8}$$

The body force **f** and the accelerations $\frac{d\,\mathbf{v}}{dt}$ are evaluated in the above equations at point **x** in the current configuration.

2.7 The Kinetic Energy Theorem

The kinetic energy K of the whole matter of the volume V is represented by the expression

$$K=\int_V \frac{1}{2}\rho\,\mathbf{v}^2 dV \tag{2.7.1}$$

The kinetic energy expression can be written in reference configuration as

$$K=\int_{V_r}\frac{1}{2}\rho_o\,\mathbf{v}^2\,dV_r \tag{2.7.2}$$

The definition of the material derivative of the integral of an extensive quantity yields

$$\frac{DK}{Dt}=\frac{dK}{dt} \tag{2.7.3}$$

The definition of material acceleration $\frac{d\,\mathbf{v}}{dt}$ given by Eq. (1.5.2) lead to the following form of Eq. (2.7.1)

$$\frac{DK}{Dt} = \int \rho \, \mathbf{v} \cdot \frac{d\,\mathbf{v}}{dt} dV \tag{2.7.4}$$

The material derivative of the kinetic energy corresponds to the opposite of the work rate of the inertia forces, which they develop in the actual movement. The inertia forces relative to the material develop their work rate in their own movement.

Let the velocity field **v*** be the actual material velocity field **v** in the virtual work rate theorem (2.5.9). Together with Eq. (2.7.4) it gives the kinetic energy theorem

$$\frac{DK}{Dt} + R_{SR}(\mathbf{v}) = R_{EF}(\mathbf{v}) \tag{2.7.5}$$

It states that in the actual movement and for any domain V, the work rate of the external forces is equal to the sum of the material derivative of the kinetic energy and of the strain work rate associated with the material strain rate.

Chapter 3

THERMODYNAMICAL BACKGROUND OF WELDING PROCESSES

3.1 Introduction

In order to derive the thermal and mechanical equations of welding processes it is necessary to consider welding as a thermodynamical process.

In such an approach the thermodynamical state of material system is characterized by the value of a finite set of state variables.

The characterization of a material leads to the characterization of its energy state, except kinetic energy. Thermostatics studies reversible and infinitely slow evolutions between two equilibrium states of homogeneous systems. The state variable necessary for the description of the evolutions throughout the first and the second laws of thermodynamics are directly observable and are linked by state equations through state functions characterizing the energy of the material system. Thermodynamics in contrast with thermostatics is the study of homogeneous systems in any evolution process. These evolutions are reversible, or not, and occur at any rate. The postulate of local state is to extend the concepts of thermostatics to these evolutions.

The postulate of local state for a homogeneous system is the following. The present state of a homogeneous system in any evolution can be characterized by the same variables as at equilibrium, and is independent of the rate of evolution.

The equilibrium fields of mechanical variables in a closed continuum, considered as a thermomechanical system are generally not homogeneous. The state of a continuum is characterized by fields of state variables. The local values of these variables characterize the state of the material particles, which constitute the continuum, and are considered as an elementary subsystem in homogeneous equilibrium.

The local state postulate for closed continua says that the elementary systems satisfy the local state postulate of a homogeneous system. The thermodynamics of closed continua is the study of material, which satisfies the extended local state postulate.

The local state postulate for closed continua is local in two aspects. It is local considering the time scale relative to the evolution rates and it refers to the space scale, which defines the dimensions of the system.

3.2 The First Law of Thermodynamics

The conservation of energy is expressed by the first law of thermodynamics. It states that the material derivative of energy $\mathcal{E}$ of the material body contained in any domain V at any time is equal to the sum of the work rate R_{EF} of the external forces acting on this body, and of the rate Q^{o} of external heat supply.

The kinetic energy $\mathcal{K}$ of the body and the internal energy E together gives the total energy of body $\mathcal{E}$. The energy has an additive character. The internal energy E can be expressed by a volume density e such that edV is the internal energy of the whole body contained in the elementary domain dV.

The energy $\mathcal{E}$ of the body volume V is expressed as

$$\mathcal{E} = \mathcal{K} + \mathsf{E} = \int_V \frac{1}{2} \rho\, v^2\, dV + \int_V e dV \tag{3.2.1}$$

where e is a volume density, not a density per mass unit.

The hypothesis of external heat supply assumes that the external heat supply is due to contact effects, with the exterior through surface a limiting the material volume V which is the external heat provided by conduction. The external heat supply is also due to external volume heat sources. The rate Q^{o} can be written as

$$Q^{o} = \int_a q da + \int_V r dV \tag{3.2.2}$$

where q is the surface rate density of heat supply by conduction. The quantity q is assumed to be a function of position vector $\mathbf{x}$, time t and outward unit normal $\mathbf{n}$ to the surface a

$$q = q\,(\mathbf{x}, t, \mathbf{n}) \tag{3.2.3}$$

In Eq. (3.2.2) the density r = r (**x**, t) is a volume rate density of the heat provided to V.

The total work rate R_{EF} (**v**) of the external forces in the whole body in volume V is

$$R_{EF}(\mathbf{v}) = \int_a \mathbf{v} \cdot \mathbf{t}\, da + \int_V \mathbf{v} \cdot \rho \mathbf{f}\, dV \tag{3.2.4}$$

then for any volume V the first law of thermodynamics reads

$$\frac{D\mathcal{E}}{Dt} = \frac{D\mathcal{K}}{Dt} + \frac{DE}{Dt} = R_{EF} + Q^{o} \tag{3.2.5}$$

Combining the first law (3.2.5) and the kinetic energy theorem (2.7.5) yields

$$\frac{DE}{Dt} = R_{SR} + Q^{o} \tag{3.2.6}$$

which expresses that the internal energy variation DE in time interval dt is due to the total strain work R_{SR}dt and the external heat supply Qdt.

3.3 The Energy Equation

The material derivative of the internal energy is obtained by letting g = e in Eq. (1.5.8) of the material derivative of a volume integral

$$\frac{DE}{Dt} = \frac{D}{Dt}\int_V e\, dV = \int_V \left[\frac{\partial e}{\partial t} + \mathrm{div}(e\,\mathbf{v})\right] dV \tag{3.3.1}$$

Making use of (3.2.2), (2.5.4), (3.1.1) and the energy balance (3.2.6) we get

$$\int_V \left[\frac{\partial e}{\partial t} + \mathrm{div}(e\,\mathbf{v}) - \mathbf{T}\mathbf{D} - r\right] dV = \int_a q\, da \tag{3.3.2}$$

Letting for f (**x**, t, **n**) = – q (**x**, t, **n**) in the relation

$$\int_V h(\mathbf{x},t)dV = -\int_a f(\mathbf{x},t,\mathbf{n})da \tag{3.3.3}$$

we get

$$q = -\mathbf{q} \cdot \mathbf{n} \tag{3.3.4}$$

where $\mathbf{q}$ is the heat flux vector. Using the expression (1.5.1) of the particulate derivative for the internal energy e, the local expression for the first law of thermodynamics in the Eulerian approach is

$$\frac{de}{dt} + e \operatorname{div} \mathbf{v} = \mathbf{T}\,\mathbf{D} + r - \operatorname{div} \mathbf{q} \tag{3.3.5}$$

The expression (3.3.5) is called the Eulerian energy equation. Multiplying Eq. (3.3.5) by dV and using Eq. (2.6.4) we have

$$\frac{d}{dt}(edV) = dr_{SR} + (r - \operatorname{div}\mathbf{q})dV \tag{3.3.6}$$

The expression (3.3.5) corresponds to a balance of internal energy for the elementary material system dV. In Eq. (3.3.6) the term d (e dV) represents the variation of internal energy of the body observed from the material particle between time t and t + dt. Equation (3.3.6) indicates that this variation is equal to the energy supplied to the open system during the same interval.

The energy supply is the sum of two terms: the elementary strain work dr_{SR}, that corresponds to the part of the external mechanical energy given to the system and not converted into kinetic energy and the external heat provided both by conduction given by the term – div $\mathbf{q}$ dt dV and by external volume heat sources given by the term r dt dV.

Define the Lagrangian vector $\mathbf{Q}$ by the relation

$$\mathbf{Q} \cdot \mathbf{N} \cdot dA = \mathbf{q} \cdot \mathbf{n} \cdot da \tag{3.3.7}$$

The expression (3.3.7) represents the heat flux $\mathbf{q} \cdot \mathbf{n} \cdot da$ throughout the oriented material surface $d\mathbf{a} = \mathbf{n} \cdot da$ in terms of the oriented surface $d\mathbf{A}_r = \mathbf{N} \cdot dA$.
By (1.3.17) we get

$$\mathbf{q} = \mathbf{F} \cdot \mathbf{Q} / J \tag{3.3.8}$$

The Lagrangian volume density of internal energy $\mathsf{E} = \mathsf{E}(\mathbf{x}, t)$ is given by

$$\mathsf{E}\, dV_r = e\, dV \tag{3.3.9}$$

By using Eq. (3.3.9)

$$\mathbf{T}\,\mathbf{D}\,dV = \mathbf{S}\frac{d\,\mathbf{E}}{dt}dV_r \qquad rdV = RdV_r \tag{3.3.10}$$

From Eq. (3.3.6) finally we get

$$\frac{dE}{dt} = \mathbf{S}\frac{d\,\mathbf{E}}{dt} + R - \operatorname{Div}\mathbf{Q} \tag{3.3.11}$$

which corresponds to the Lagrangian formulation of the Eulerian energy equation (3.3.5).

3.4 The Second Law of Thermodynamics

The conservation of energy is expressed by the first law. The second law is of a different kind. It states that the quality of energy can only deteriorate. The quantity of energy transformed to mechanical work can only decrease irreversibly. In the second law a new physical quantity is introduced, entropy, which represents a measure of this deterioration and which can increase when considering an isolated system. In a system that is no longer isolated, the second law defines a lower bound to the entropy increase, which takes into account the external entropy supply. The latter is defined in term of a new variable, the temperature. The second law can be formulated as below.

The material derivative of a thermodynamic function S, called entropy attached to any material system V is equal or superior to the rate of entropy externally supplied to it. The external entropy rate supply can be defined in terms of a universal scale of absolute temperature denoted by θ_r and positively defined.

The external entropy rate is then defined as the ratio of the heat supply rate and the absolute temperature at which the heat is provided to the considered subsystem. We denote by s the entropy volume density, such that the quantity sdV represents the entropy of all the matter presently contained in the open elementary system dV.

The total entropy S of all the matter contained in V is

$$\mathsf{S} = \int_V s\,dV \tag{3.4.1}$$

The second law reads

$$\frac{D\mathsf{S}}{Dt} \geq -\int_a \frac{\mathbf{q}\cdot\mathbf{n}}{T}\,da + \int_V \frac{r}{\theta_r}\,dV \tag{3.4.2}$$

The right-hand side of inequality (3.4.2) represents the rate of external entropy supply. This external rate is composed of both the entropy influx associated with the heat provided by conduction through surface a enclosing the considered material volume V, and of the volume entropy rate associated with the external heat sources distributed within the same volume.

Eq. (3.4.2) implies that the internal entropy production rate cannot be negative in real evolutions. The entropy S of a material system V at $\mathbf{q} = 0$ and $r = 0$ cannot spontaneously decrease.

The material derivative of a volume integral in the Eulerian expression for $g = s$ reads

$$\frac{D\mathsf{S}}{Dt} = \frac{D}{Dt}\int_V s\,dV = \int_V \left[\frac{\partial s}{\partial t} + \operatorname{div}\left(s\,\mathbf{v}\right)\right] dV \tag{3.4.3}$$

By Eqs. (3.4.2) and (3.4.3) it follows that the volume integral must be non-negative for any system V,

$$\frac{\partial s}{\partial t} + \operatorname{div}\left(s\,\mathbf{v}\right) + \operatorname{div}\frac{\mathbf{q}}{\theta_r} - \frac{r}{\theta_r} \geq 0 \tag{3.4.4}$$

where the surface integral has been transformed to the volume integral. The expression (3.4.4) can be rewritten as

$$\frac{\partial s}{\partial t} + s\operatorname{div}\mathbf{v} + \operatorname{div}\frac{\mathbf{q}}{\theta_r} - \frac{r}{\theta_r} \geq 0 \tag{3.4.5}$$

Multiplying Eq. (3.4.5) by dV, the above inequality becomes

$$\frac{d}{dt}(s\,dV)+\left(\operatorname{div}\frac{\mathbf{q}}{\theta_r}-\frac{r}{\theta_r}\right)dV \geq 0 \tag{3.4.6}$$

In Eq. (3.4.6) the term $\frac{d}{dt}(sdV)$ represents the variation in entropy of this open system, during the infinitesimal time interval dt observed from any material point.

By energy equation (3.3.5), the fundamental inequality (3.4.5) can be written as

$$\mathbf{T\,D}+\theta_r\frac{ds}{dt}-\frac{de}{dt}-(e-\theta_r\,s)\operatorname{div}\mathbf{v}-\frac{\mathbf{q}}{\theta_r}\cdot\operatorname{grad}\theta_r \geq 0 \tag{3.4.7}$$

Now we define the free volume energy of the open system dV

$$\psi = e-\theta_r\,s \tag{3.4.8}$$

By Eqs. (3.4.8) and (3.4.7) we get

$$\mathbf{T\,D}+s\frac{d\theta_r}{dt}-\frac{d\psi}{dt}-\psi\operatorname{div}\mathbf{v}-\frac{\mathbf{q}}{\theta_r}\cdot\operatorname{grad}\theta_r \geq 0 \tag{3.4.9}$$

The Eq. (3.4.9) is the fundamental inequality or so-called Clausius-Duhem inequality. It corresponds to a Eulerian approach.

The Lagrangian approach of the fundamental inequality leads to expressing DS/Dt in terms of the Lagrangian entropy density S defined by

$$S\,dV_r = s\,dV \tag{3.4.10}$$

By transport formula (3.3.7) and (3.3.10) together with a Lagrangian expression of the material derivative of the integral of an extensive variable, the inequality (3.4.2) can be rewritten as

$$\frac{DS}{Dt}=\frac{D}{Dt}\int_{V_r}S\,dV_r=\int_{V_r}\frac{dS}{dt}dV_r \geq -\int\frac{\mathbf{Q}\cdot\mathbf{N}}{\theta_r}dA_r+\int_{V_r}\frac{R}{\theta_r}dV_r \tag{3.4.11}$$

The expression (3.4.11), which holds for any domain V_r can be written in Lagrangian formulation of the Eulerian inequality in the form

$$\frac{dS}{dt} + \text{Div}\frac{\mathbf{Q}}{\theta_r} - \frac{R}{\theta_r} \geq 0 \tag{3.4.12}$$

By the Lagrangian energy equation (3.3.11) and the positiveness at absolute temperature θ_r, from the inequality (3.4.12) it follows that

$$\mathbf{S}\frac{d\mathbf{E}}{dt} + \theta_r \frac{dS}{dt} - \frac{dE}{dt} - \frac{\mathbf{Q}}{\theta_r} \cdot \text{Grad}\,\theta_r \geq 0 \tag{3.4.13}$$

We define the free Lagrangian energy density Ψ as

$$\Psi\, dV_r = \psi\, dV \qquad \Psi = E - \theta_r S \tag{3.4.14}$$

By Eqs. (3.4.13) and (3.4.14) we get

$$\mathbf{S}\frac{d\mathbf{E}}{dt} - S\frac{d\theta_r}{dt} - \frac{d\Psi}{dt} - \frac{\mathbf{Q}}{\theta_r} \cdot \text{Grad}\,\theta_r \geq 0 \tag{3.4.15}$$

The expression (3.4.15) is the Lagrangian formulation of the Clausius-Duhem inequality.

3.5 Dissipations

The left-hand side of Eq. (3.4.15) which will be noted Π is the dissipation per unit of initial volume dV_r. The second law requires the dissipation Π and the associated internal entropy production Π/θ_r to be non-negative. The dissipation Π is the sum of two terms

$$\Pi = \Pi_1 + \Pi_2 \tag{3.5.1}$$

where

$$\Pi_1 = \mathbf{S}\frac{d\mathbf{E}}{dt} - S\frac{d\theta_r}{dt} - \frac{d\Psi}{dt} \tag{3.5.2}$$

The local state postulate assumes that all the quantities appearing in Eq. (3.5.2) depend only on the state variables characterizing the free energy ΨdV_r of the open elementary system dV_r.

The second dissipation Π_2 is defined as

$$\Pi_2 = -\frac{\mathbf{Q}}{\theta_r} \cdot \text{Grad}\, \theta_r \tag{3.5.3}$$

and is called the thermal dissipation associated with heat conduction. By definition (3.4.14) of Ψ and definition of Π_1 (3.5.2), the Lagrangian energy equation (3.3.11) is rewritten as

$$\theta_r \frac{dS}{dt} = R - \text{Div}\, \mathbf{Q} + \Pi_1 \tag{3.5.4}$$

The expression (3.5.4) is called the Lagrangian thermal equation. Using Eqs. (3.5.3) and (3.5.4) we get

$$\frac{d}{dt}\left(S dV_r\right) = \left[\frac{R}{\theta_r} - \text{Div}\left(\frac{\mathbf{Q}}{\theta_r}\right)\right] dV_r + \frac{\Pi}{\theta_r} dV_r \tag{3.5.5}$$

The thermal equation (3.5.4) or equivalently Eq. (3.5.5) corresponds to a balance in entropy for the elementary system dV_r. The term $d(SdV_r)$ is the entropy variation, during the time interval dt, observed from the point of the open system dV_r.

The internal source of entropy (Π/θ_r) dt dV_r is the sum of the production (Π_2/θ_r) dt dV_r associated with heat conduction and the production (Π_1/θ_r) dt dV_r associated with the mechanical dissipation Π_1 dV_r dt. As it will be seen later, Π_2 dV_r dt corresponds to a mechanical energy converted into heat by mechanical dissipation. Thus, the mechanical dissipation Π_1 appears as a heat source in thermal equation (3.5.4).

A Lagrangian approach to thermal equation is introduced below. If Π_i is the Eulerian dissipation volume densities, then

$$\pi_i\, dV = \Pi_i\, dV_r \qquad J\, \pi_i = \Pi_i \tag{3.5.6}$$

The respective expressions can be written as

$$\begin{aligned} \pi_1 &= \mathbf{T}\,\mathbf{D} - s\frac{d\theta_r}{dt} - \frac{d\Psi}{dt} - \Psi \,\text{div}\, \mathbf{v} \\ \pi_2 &= -\frac{\mathbf{q}}{\theta_r} \cdot \text{grad}\, \theta_r \end{aligned} \tag{3.5.7}$$

By Eqs. (3.5.1) and (3.5.6), for the densities π_i the following relation holds

$$\pi_1 + \pi_2 \geq 0 \tag{3.5.8}$$

The Eulerian thermal equation takes the form

$$\theta_r \left[\frac{ds}{dt} + s \operatorname{div} \mathbf{v} \right] = r - \operatorname{div} \mathbf{q} + \pi_1 \tag{3.5.9}$$

3.6 Equations of State for an Elementary System

For any evolution of the continuum the energy states of the elementary systems depend on the same state variables as at equilibrium. The free energy volume density Ψ depends locally on the state variables, but not on their rates nor on their spatial gradients. We assume that it is also true for the Piola-Kirchhoff stress tensor **S**. In Eq. (3.5.1) the dissipation Π_2 depends on Grad T. Let Grad T = 0 ($\Pi_2 = 0$), then the non-negativeness of intrinsic dissipation Π_1 is derived independently of the non-negativeness of total dissipation Π

$$\Pi_1 = \mathbf{S} \frac{d\mathbf{E}}{dt} - S \frac{d\theta_r}{dt} - \frac{d\Psi}{dt} \geq 0 \tag{3.6.1}$$

The expression (3.6.1) is derived from the second law and the local state postulate.

In Eq. (3.6.1) the energy **S** d**E** represents the energy supplied to the elementary system dV_r, not in the form of heat and not converted into kinetic energy during the time interval dt. It is actually supplied if positive, extracted if negative. The energy $d\Psi + Sd\theta_r = dE - \theta_r dS$ is the part of the previous energy that the system effectively stores in the time interval dt in any other form but heat. It is actually stored if positive, extracted if negative.

The Lagrangian expression of the particulate derivatives used here do not involve gradients, contrary to the corresponding Eulerian expression, which has importance in this reasoning.

The free energy Ψ can be written as

$$\Psi = \Psi\,(\theta_r, E_{\alpha\beta}, m_1, \ldots, m_n) \tag{3.6.2}$$

based on the local state postulate.

The variables θ_r, $E_{\alpha\beta}$, m_1, …, m_n constitute a set of state variables characterizing the state of the open system dV_r. By the local state postulate, these state variables are macroscopic variables.

The non-negativeness of the intrinsic dissipation expressed by (3.6.1) gives

$$\left(\mathbf{S}-\frac{\partial \Psi}{\partial \mathbf{E}}\right)\frac{d\mathbf{E}}{dt}-\frac{\partial \Psi}{\partial \mathbf{m}}\cdot\frac{d\mathbf{m}}{dt}-\left(S+\frac{\partial \Psi}{\partial \theta_r}\right)\frac{d\theta_r}{dt}\geq 0 \tag{3.6.3}$$

In Eq. (3.6.3) $\partial\Psi/\partial\mathbf{E}$ stands for the tensor of components $\partial\Psi/\partial E_{\alpha\beta}$ and

$$\frac{\partial \Psi}{\partial \mathbf{m}}\cdot\frac{d\mathbf{m}}{dt}=\frac{\partial \Psi}{\partial \mathbf{m}}\cdot\dot{\mathbf{m}}=\frac{\partial \Psi}{\partial m_I}\cdot\dot{m}_I \tag{3.6.4}$$

where

$$\dot{\mathbf{m}}=\frac{d\mathbf{m}}{dt} \tag{3.6.5}$$

The inequality (3.6.3) is relative only to this elementary open system dV_r under consideration.

We state that a set of state variables characterizing an elementary system is normal if variations of this particular state can occur independently of the variations of the other variables of the set.

The hypothesis introduced by Helmholtz states that it is possible to find a set of variables, which is normal with respect to absolute temperature θ_r. In Eq. (3.6.3) with such a set of variables, real evolution can occur with arbitrary variations $d\theta_r$ and zero variations for the other variables of the set. From Eq. (3.6.3) it follows that

$$S=-\frac{\partial \Psi}{\partial \theta_r} \tag{3.6.6}$$

because the inequality (3.6.3) must remain satisfied for these particular real evolutions, and assuming that the present value of entropy S is independent of the temperature rate $d\theta_r/dt$.

Assume that the variable set is normal with respect to the state variables $E_{\alpha\beta}$, and that the present value of the Piola-Kirchoff stress tensor **S** is independent of the rates $d\mathbf{E}/dt$.

Then we have

$$\mathbf{S} = \frac{\partial \Psi}{\partial \mathbf{E}} \tag{3.6.7}$$

Equation (3.6.7) associates the state variables θ_r, $E_{\alpha\beta}$ with their dual thermodynamic variables –S and **S**. They are called the state equations of the system. They still hold for a system of evolution in holding at equilibrium.

3.7 The Heat Conduction Law

By the second law of thermodynamics and the local state postulate we get the relation

$$\pi_1 + \pi_2 \geq 0 \qquad \pi_1 \geq 0 \tag{3.7.1}$$

where the Eulerian dissipations π_1 and π_2 are the intrinsic dissipation and the dissipations associated with transport phenomena, per unit volume dV in the current configuration expressed by (3.5.6). The internal entropy production rate π_2/θ_r is due to the assembly of adjacent elementary systems that ensures the continuity of the medium, contrary to the intrinsic internal entropy production rate π_1/θ_r related to the elementary system, which is considered independently of the other system.

The hypothesis of dissipation decoupling which is more restrictive than the above assumes

$$\pi_1 \geq 0 \qquad \pi_2 \geq 0 \tag{3.7.2}$$

The heat conduction law will be formulated on the basis of hypothesis (3.7.2).

By (3.6.2) and (3.6.7) the dissipation Π_1 fulfills the expression

$$\Pi_1 = -\frac{\partial \Psi}{\partial \mathbf{m}} \cdot \dot{\mathbf{m}} \geq 0 \tag{3.7.3}$$

The above equation can be written in the form

$$\Pi_1 = \mathbf{B}_{\dot{m}} \cdot \dot{\mathbf{m}} \geq 0 \qquad \mathbf{B}_{\dot{m}} = -\frac{\partial \Psi}{\partial \dot{\mathbf{m}}} \tag{3.7.4}$$

The rates $\dot{\mathbf{m}}_I$ of the internal variables are associated with the thermodynamical forces $\mathbf{B}\,\dot{\mathbf{m}}_I$.

The hypothesis of normality of a dissipative mechanism consists of introducing the existence of both a set of internal variables $\mathbf{m}_I$ and a function H of their rates

$$\mathbf{B}_{\dot{\mathbf{m}}} = -\frac{\partial H}{\partial \dot{\mathbf{m}}} \tag{3.7.5}$$

The function H is called the dissipation potential.
The hypothesis (3.7.2) requires the non-negativeness of thermal dissipation π_2

$$\mathbf{B}_{q/\theta_r} \cdot \frac{\mathbf{q}}{\theta_r} \geq 0 \qquad \mathbf{B}_{q/\theta_r} = -\operatorname{grad}\theta_r \tag{3.7.6}$$

From Eq. (3.7.6) the entropy vector $\mathbf{q}/\theta_r$ and the thermodynamic force – grad θ_r are associated as in Eq. (3.7.6). The decoupling hypothesis (3.7.2) and the resulting inequality (3.7.6) states that the heat flows from high temperatures. The positiveness of the associated internal entropy production rate π_2/θ_r expresses that it is not-less evident to find a cold source to extract efficient mechanical work from this heat.

Let H_2 ($\mathbf{q}/\theta_r$) be the dissipation potential. Assume the normality of the associated dissipative mechanism. From Eq. (3.7.6) we get the heat conduction law

$$-\operatorname{grad}\theta_r = \frac{\partial H_2}{\partial(\mathbf{q}/\theta_r)} \tag{3.7.7}$$

If we define H_2 as

$$H_2\left(\frac{\mathbf{q}}{\theta_r}\right) = \frac{1}{2\theta_r}\,\mathbf{q}\cdot\mathbf{k}^{-1}\cdot\mathbf{q} \tag{3.7.8}$$

where $\mathbf{k}$ is a symmetric tensor, Eqs. (3.7.7) and (3.7.8) give the linear heat conduction law called the Fourier law

$$\mathbf{q} = -\mathbf{k}\operatorname{grad}\theta_r \tag{3.7.9}$$

where $\mathbf{k}$ is the so-called thermal conductivity tensor, relative to the current configuration. Since H_2 defined by Eq. (3.7.8) is a quadratic function, the corresponding irreversible process is linear and the thermal dissipation associated with heat transport is $\Pi_2 = H_2$ per unit volume dV.

In Lagrangian approach we have

$$H_2\left(\frac{\mathbf{q}}{\theta_r}\right)dV = H_2\left(\frac{\mathbf{Q}}{\theta_r}\right)dV_r \tag{3.7.10}$$

$$H_2\left(\frac{\mathbf{Q}}{\theta_r}\right) = \frac{1}{2\theta_r}\mathbf{Q}\cdot\mathbf{K}^{-1}\cdot\mathbf{Q} \tag{3.7.11}$$

$$-\,\mathrm{Grad}\,\theta_r = \frac{\partial H_2}{\partial\left(\frac{\mathbf{Q}}{\theta_r}\right)} \tag{3.7.12}$$

and finally we get

$$\mathbf{Q} = -\,\mathbf{K}\cdot\mathrm{Grad}\,\theta_r \tag{3.7.13}$$

where

$$\mathbf{K} = J\,\mathbf{F}^{-1}\cdot\mathbf{k}\cdot\left(\mathbf{F}^{-1}\right)^T \tag{3.7.14}$$

For isotropic material in reference configuration the tensor **K** is written as

$$\mathbf{K} = K\,\mathbf{1} \qquad \mathbf{k} = (K/J)\mathbf{F}\cdot^{T}\mathbf{F} \tag{3.7.15}$$

3.8 Thermal Behaviour

The thermal behaviour of material is defined by assuming the dissipation being equal to zero in any evolution, and thus by the absence of internal variables. The constitutive equations of the thermal system reduce the state equation to

$$S = -\frac{\partial\Psi}{\partial\theta_r} \tag{3.8.1}$$

where the energy Ψ depends on the external variable θ_r

$$\Psi = \Psi(\theta_r) \tag{3.8.2}$$

The expression (3.8.2) has to be specified to get the constitutive equations. The linearization presented here assumes small temperature variations $\theta = \theta_r - \theta_r^o$. Under this assumption the expression of the free energy $\Psi = \Psi(\theta_r)$ as a second-order expansion with regard to argument θ is assumed as

$$\Psi = -S_o\theta - \frac{1}{2}b\theta^2 \tag{3.8.3}$$

By Eq. (3.8.3) we get

$$S = S_o + b\theta \tag{3.8.4}$$

The time differentiation of Eq. (3.8.4) gives

$$\theta_r^o \frac{dS}{dt} = \theta_r^o b \frac{d\theta}{dt} \tag{3.8.5}$$

In the physical linearization limit, the thermal Eq. (3.5.4) reads

$$\theta_r^o \frac{dS}{dt} = R - \mathrm{Div}\,\mathbf{Q} \tag{3.8.6}$$

In Eq. (3.8.6) R – Div **Q** represents the external rate of heat supply to the elementary open system and $C_\Delta = \theta_r^o b$ is the volume heat capacity per unit of initial volume, so that the heat needed to produce a temperature variation θ in an iso-deformation (**E** = 0) is equal to $C_\Delta\theta$.

3.9 Heat Transfer in Cartesian Coordinates

In the case of infinitesimal transformation in Cartesian coordinates the point of the space is described by the position vector $\mathbf{x} = \mathbf{x}(x_1, x_2, x_3)$. Under the assumption of infinitesimal transformation the reference configuration is equal to the current configuration, and the following relations hold $\mathbf{x} = \mathbf{x}$, $\mathbf{F} = \mathbf{1}$, $J = 1$ and $\mathbf{r} = \mathbf{R}$, $\mathbf{q} = \mathbf{Q}$. Moreover for simplicity let $\theta_r^o = 0$.
By Eqs. (3.8.5) and (3.8.6) we get

$$C_\Delta \frac{d\theta}{dt} = r - \mathrm{Div}\,\mathbf{q} \tag{3.9.1}$$

By Eq. (3.7.6) we get from (3.9.1)

$$C_\Delta \frac{d\theta}{dt} = r + \operatorname{div}(\mathbf{k}\,\operatorname{grad}\theta) \qquad C_\Delta \frac{d\theta}{dt} = r + \frac{\partial}{\partial x_i}\left(k_{ij}\frac{\partial\theta}{\partial x_j}\right) \tag{3.9.2}$$

Using the definition of particulate derivative, Eq. (3.9.2) becomes

$$C_\Delta\left(\frac{\partial\theta}{\partial t} + \mathbf{v}\,\operatorname{grad}\theta\right) = r + \operatorname{div}(\mathbf{k}\,\operatorname{grad}\theta)$$
$$C_\Delta\left(\frac{\partial\theta}{\partial t} + v_i\frac{\partial\theta}{\partial x_i}\right) = r + \frac{\partial}{\partial x_i}\left(k_{ij}\frac{\partial\theta}{\partial x_j}\right) \tag{3.9.3}$$

If $\mathbf{v} = 0$, then we get the following form of heat transfer equation

$$C_\Delta \frac{\partial\theta}{\partial t} = r + \operatorname{div}(\mathbf{k}\,\operatorname{grad}\theta) \tag{3.9.4}$$

3.10 Heat Convection

A frequently encountered case of practical significance in welding is the heat exchange between a material wall and an adjacent gas or liquid. Heat exchange in fluids takes place by convection, but near a wall there exists a very thin layer in which heat exchange takes place by conduction. In the case where heat exchange is stationary, the heat is transferred from the wall towards the center of the fluid. If the intensity of heat transferred is higher, then the drop in temperature per unit length in a direction perpendicular to the wall is lower. Near the wall one observes a significant drop of temperature because in the thin boundary layer conduction plays a decisive role and heat exchange is less intensive in the boundary layer than in areas remote from the wall, where convection also takes place.

The phenomenon described above is known as heat convection. Mathematically it is described by the Newton equation

$$q = h\,(\theta_w - \theta_f) \tag{3.10.1}$$

where θ_w is the wall temperature, θ_f is the fluid temperature at a sufficiently great distance from the wall, and the method of determination of θ_f is usually precisely laid down. The magnitude h determining the heat exchange intensity is called the heat convection coefficient.

3.11 Heat Radiaton

A black body which is essential to radiation theory is a hypothetical body which absorbs all radiant energy falling onto it, transmitting and reflecting nothing. Heat radiation occurs in accordance with the Stefan-Boltzmann law, which states that the energy radiated by a black body is proportional to the fourth power of the absolute temperature of that body. Mathematically this law is expressed by the formula

$$q = C_o \left[\frac{\theta_r}{100}\right]^4 \tag{3.11.1}$$

where C_o is the so-called radiation coefficient of the black body and θ_r is the absolute temperature. The heat radiated through the surface A per unit time is

$$q_h = C_o A \left[\frac{\theta_r}{100}\right]^4 \tag{3.11.2}$$

Real bodies are not black bodies and at a given temperature will radiate less energy than a black body. If the ratio of the energy radiated by the real body to the energy radiated by the black body in the same conditions does not depend on radiation wavelength, then this body is called a gray body. The heat exchange between gray bodies is described by the equation

$$q_{1-2} = C_o A_1 \phi_{1-2} \left[\left(\frac{\theta_r^1}{100}\right)^4 - \left(\frac{\theta_r^2}{100}\right)^4\right] \tag{3.11.3}$$

where θ_r^1 and θ_r^2 are the absolute temperatures of the bodies radiating the heat, A_1 is the surface of the body at temperature θ_r^1, and ϕ_{1-2} is the coefficient taking into consideration the deviation of the properties of the analyzed body from the properties of the black body and the geometrical system of the two bodies.

Heat exchange based on pure conduction, convection or radiation holds during welding very rarely. These three fundamental kinds of heat exchange normally appear in various combinations. A common case is the exchange of heat through a solid wall by a combination of radiation and convection. In this case a substitute coefficient of heat exchange by radiation h_r is introduced which is defined as follows

$$h_r = \frac{q_{1-2}}{A_1\left(\theta_r^1 - \theta_r^0\right)} \tag{3.11.4}$$

where q_{1-2} is the heat exchanged by radiation, given by Eq. (3.11.3), θ_r^1 is the wall temperature, and θ_r^0 is the reference temperature. The reference temperature θ_r^0 does not have to be equal to the temperature θ_r^2 appearing in Eq. (3.11.3), and in the case of convection and radiation one puts it equal to the fluid temperature θ_r^f. The expression (3.11.4) can then be rewritten in the form

$$h_r = \frac{C_o \phi_{1-2}\left[\left(\frac{\theta_r^1}{100}\right)^4 - \left(\frac{\theta_r^2}{100}\right)^4\right]}{\theta_r^1 - \theta_r^f} \tag{3.11.5}$$

Heat exchange by both convection and radiation can be described by the relation

$$q = \left(h + h_r\right)\left(\theta_r^w - \theta_r^f\right) \tag{3.11.6}$$

where h_r is the heat radiation coefficient described by Eq. (3.11.4) or (3.11.5), and θ_r^w and θ_r^f are wall and fluid temperatures respectively.

3.12 Initial and Boundary Conditions

The initial and boundary conditions are necessary to solve the heat condition equation. The initial condition prescribes the temperature at time $t = 0$, that is

$$\theta\,(\mathbf{x}, 0) = f(\mathbf{x}) \tag{3.12.1}$$

The boundary conditions describe the heat exchange at the boundary of the body and are given by one of three possible cases.

1. The temperature distribution on the boundary of the body at any time t

$$\theta_r^w(t) = g(t) \tag{3.12.2}$$

where $\theta_r^w(t)$ is the body surface temperature. This condition is called a boundary condition of the first kind.

2. The heat flux is determined at each point of the body surface

$$q_w\,(t) = h\,(t) \tag{3.12.3}$$

This condition is called a boundary condition of the second kind.

3. The temperature of the surrounding medium and the relation describing the heat exchange between the heat-conducting material and the surroundings are known. The heat exchange between the heat-conducting body and its surroundings takes place by convection, radiation or by both of these phenomena and is most conveniently described by the Newton equation (3.10.1)

The Newton equation written for the surface element da

$$dq_h = \alpha\left(\theta_r^w - \theta_r^f\right)da \tag{3.12.4}$$

shows the amount of heat exchanged by the element with the surroundings. On the other hand the same amount of heat has to be conducted at the boundary of the body, i.e.

$$dq_h = -\,k\,(\text{grad}\,\theta)_w\,da \tag{3.12.5}$$

where $(\text{grad}\,\theta)_w$ denotes the magnitude of the temperature gradient between the boundary of the body and the surroundings. Comparison of the above two expressions for dq_h gives

$$(\text{grad}\,\theta)_w = -\frac{\alpha}{k}\left(\theta_w - \theta_f\right) \tag{3.12.6}$$

The above condition is called the boundary condition of the third kind. In the mathematical theory of heat conduction the ratio α/k is often denoted by h and is called the heat exchange coefficient.

A fourth kind of boundary condition covers heat exchange with surroundings by conduction. In heat exchange with surroundings by conduction the material surface temperature $\theta_w^{'}$ and the surroundings material temperature $\theta_w^{''}$ are identical

$$\theta_w^{'}(t) = \theta_w^{''}(t) \tag{3.12.7}$$

Moreover magnitudes of heat fluxes on the surface separating the materials considered are identical

$$-k'\left(\frac{\partial\theta}{\partial n}\right)_w' = -k''\left(\frac{\partial\theta}{\partial n}\right)_w'' \qquad (3.12.8)$$

Chapter 4

MOTION OF FLUIDS

4.1 Viscous Fluids

In welding processes a material is analyzed either in solid or fluid state. If under the action of forces the deformation of the body increases continuously and indefinitely with time we say that a material flows. In the case of a purely viscous material, a stress is only generated if the amount of strain is changing, the stress generated by strain being considered to relax instantaneously.

Euler's equation of motion of an ideal fluid can be written in the form

$$\frac{\partial}{\partial t}(\rho\,\mathbf{v}) = -\operatorname{div}\mathbf{\Pi} \tag{4.1.1}$$

where

$$\mathbf{\Pi} = P\,\mathbf{I} + \rho\,\mathbf{v}\otimes\mathbf{v} \tag{4.1.2}$$

is the momentum flux density tensor and P is the thermodynamic pressure. The momentum flux given by Eq. (4.1.2) represents a completely reversible transfer of momentum, due to the mechanical transport of the different particles of fluid from one point to another and to the pressure forces acting on the fluid.

The viscosity (i.e. internal friction) is due to another, irreversible transfer of momentum from points where the velocity is larger to those where it is small. The equation of motion of a viscous fluid may be obtained by adding to the ideal momentum flux of Eq. (4.1.1) a term $-\mathbf{T}_E$ which gives the irreversible transfer of momentum in the fluid. Thus the momentum flux density tensor in a viscous fluid is written in the form

$$\begin{aligned}\mathbf{\Pi} &= \mathrm{P}\,\mathbf{I} + \rho\,\mathbf{v}\otimes\mathbf{v} - \mathbf{T}_{\mathrm{E}} \\ &= -\,\mathbf{T} + \rho\,\mathbf{v}\otimes\mathbf{v}\end{aligned} \tag{4.1.3}$$

The tensor

$$\mathbf{T} = -\mathrm{P}\,\mathbf{I} + \mathbf{T}_{\mathrm{E}} \tag{4.1.4}$$

is the stress tensor and the tensor $\mathbf{T}_{\mathrm{E}}$ is called the extra stress.
For the no-flow condition the stress must be of the form

$$[\mathbf{T}]_0 = -\overline{\mathrm{P}}_0\,\mathbf{I} \tag{4.1.5}$$

where $\overline{\mathrm{P}}_0$ is the hydrostatic pressure. For an ideal nonviscous fluid, $\mathrm{P} = \overline{\mathrm{P}}_0$ where P is the thermodynamic pressure.

A definition of a purely viscous fluid was proposed by Stokes. It was based on the assumption that the difference between the stress in a deforming fluid and the static equilibrium stress given by Eq. (4.1.5), i.e., the extra stress $\mathbf{T}_{\mathrm{E}}$ depends only on the relative motion of the fluid.
Thus, the constitutive assumption of Stokes may be expressed in the form

$$\mathbf{T}_{\mathrm{E}} = \mathrm{f}(\mathbf{D}) \tag{4.1.6}$$

that is

$$\mathbf{T} = -\mathrm{P}\,\mathbf{I} + \mathrm{f}(\mathbf{D}) \qquad \mathrm{f}(\mathbf{0}) = \mathbf{0} \tag{4.1.7}$$

where P is the thermodynamic pressure.

When the function f is linear, the fluid is referred to as a Newtonian fluid and the constitutive equation can be expressed in the form

$$\mathbf{T} = -\mathrm{P}\,\mathbf{I} + \mathbf{A}\mathbf{D} \tag{4.1.8}$$

Eq. (4.1.8) can be transformed to the form

$$\mathbf{T} = -\mathrm{P}\,\mathbf{I} + \mathrm{A}_{(1)}(\mathrm{tr}\,\mathbf{D})\mathbf{I} + 2\mathrm{A}_{(2)}\,\mathbf{D} \tag{4.1.9}$$

where $\mathrm{A}_{(1)}$ and $\mathrm{A}_{(2)}$ are scalars and (tr$\mathbf{D}$) is the first invariant of the rate of strain tensor.

Eq. (4.1.9) can be expressed in the form

$$\mathbf{T} = -P\,\mathbf{I} + \mu(\mathrm{tr}\,\mathbf{D})\mathbf{I} + 2\eta\,\mathbf{D} \tag{4.1.10}$$

where $A_{(1)} = \mu = \text{const}$ and $A_{(2)} = \eta = \text{const}$.

Eq. (4.1.10) is the Navier-Poisson law. It is evident from Eq. (4.1.10) that two constants η and μ characteristic the material properties, are required to define the properties of the Newtonian fluid. Since the shearing components of $\mathbf{T}_E$ are given by the term $2\eta D_{ij}$ $(i \neq j)$, the coefficient η is called the shear, or dynamic viscosity.

Eq. (4.1.10) can be expressed in terms of the deviators

$$\mathbf{T}' = \mathbf{T} - \frac{1}{3}(\mathrm{tr}\,\mathbf{T})\mathbf{I} \tag{4.1.11}$$

$$\mathbf{D}' = \mathbf{D} - \frac{1}{3}(\mathrm{tr}\,\mathbf{D})\mathbf{I} \tag{4.1.12}$$

to give the constitutive equation for a Newtonian fluid in the form

$$\mathbf{T}' = \left(\overline{P} - P\right)\mathbf{T} + \left(\mu + \frac{2}{3}\eta\right)(\mathrm{tr}\,\mathbf{D})\mathbf{I} + 2\eta\,\mathbf{D}' \tag{4.1.13}$$

where

$$\overline{P} = -\frac{1}{3}(\mathrm{tr}\,\mathbf{T}) \tag{4.1.14}$$

is the mean pressure. Since $\mathrm{tr}\,\mathbf{T}' = 0$, and $\mathrm{tr}\,\mathbf{D}' = 0$, Eq. (4.1.13) gives

$$\overline{P} - P + \left(\mu + \frac{2}{3}\eta\right)(\mathrm{tr}\,\mathbf{D}) = 0 \tag{4.1.15}$$

The condition of Eq. (4.1.15) together with the equation of continuity expressed in the form

$$\mathrm{tr}\,\mathbf{D} = \mathrm{div}\,\mathbf{v} = -\frac{1}{\rho}\frac{D\rho}{Dt} \tag{4.1.16}$$

reduces Eq. (4.1.13) to the relation

$$\mathbf{T}' = 2\eta\,\mathbf{D}' \tag{4.1.17}$$

$$\overline{P} = P - \xi\left(\operatorname{tr}\mathbf{D}\right) = P + \xi\frac{1}{\rho}\frac{D\rho}{Dt} \tag{4.1.18}$$

when $\xi = \mu + \frac{2}{3}\eta$ is the bulk viscosity.

From Eq. (4.1.18) it is evident that the mean pressure $\overline{P}$ equals the thermodynamic pressure P if the following two conditions is satisfied

$$\operatorname{tr}\mathbf{D} = 0 \tag{4.1.19}$$

or the bulk viscosity

$$\xi = \mu + \frac{2}{3}\eta = 0 \tag{4.1.20}$$

which is called the Stokes condition.
For an incompressible fluid Eq. (4.1.10) reduces to

$$\mathbf{T} = -P\,\mathbf{I} + 2\eta\,\mathbf{D} \tag{4.1.21}$$

4.2 Navier-Stokes Equation

A fluid obeying Fourier's law of heat conduction satisfies the thermal equation

$$C_\Delta\left(\frac{\partial\theta}{\partial t} + \mathbf{v}\nabla\theta\right) = \nabla\left(k\nabla\theta\right) + r \tag{4.2.1}$$

Density changes are allowed to occur in the fluid in response to changes in the temperature according to the relation

$$\rho = \rho_0\left[\left(1 - \beta\left(\theta - \theta_0\right)\right)\right] \tag{4.2.2}$$

where β is the coefficient of volumetric thermal expansion and subscript o refers to the reference conditions.

For the incompressible case the conservation of mass equation is written in the form

$$\operatorname{div} \mathbf{v} = 0 \tag{4.2.3}$$

If g is the acceleration due to gravity, then

$$f = g\beta(\theta - \theta_0) \tag{4.2.4}$$

and we get the equation of motion in the form

$$\rho \frac{d\mathbf{v}}{dt} = \operatorname{div} \mathbf{T} - \rho g \beta(\theta - \theta_0) \tag{4.2.5}$$

To complete the formulation of the boundary-value problem, suitable boundary conditions for the dependent variables are required. For the hydrodynamic part of the problem either velocity components or the total surface stress (or traction) must be specified on the boundary of the fluid region. The thermal part of the problem requires a temperature or heat flux to be specified on all parts of the boundary. Symbolically, these conditions are expressed by

$$v_i = f_i(s) \quad \text{on } a_v \tag{4.2.6}$$

$$t_i = \tau_{ij}(s) n_j(s) \quad \text{on } a_t \qquad a = a_v \cup a_t \tag{4.2.7}$$

In Eqs. (4.2.6) and (4.2.7) s denotes a generic point on the boundary, and n_j are the components of the outward unit normal to the boundary.

Chapter 5

THERMOMECHANICAL BEHAVIOUR

5.1 Thermoelasticity

In the weld and base material complex strains occur being the result of welding heat source acting on the welded structure. The deformation of welded structure is thermo-elastic or thermo-elastic-plastic. Thermo-mechanical coupling in thermo-elastic material is described by the following state equations

$$\mathbf{S} = \frac{\partial \Psi}{\partial \mathbf{E}} \qquad S = -\frac{\partial \Psi}{\partial \theta_r} \tag{5.1.1}$$

In thermo-elasticity the free energy Ψ depends on temperature θ_r and strain **E**

$$\Psi = \Psi\,(\theta_r, \mathbf{E}) \tag{5.1.2}$$

Assume small temperature variations $\theta = \theta_r - \theta_r^0$. The free energy is expressed in the form

$$\Psi = \mathbf{S}^{\circ}\mathbf{E} - S_0\theta + \frac{1}{2}\mathbf{E}\,\mathbf{C}\,\mathbf{E} - \theta\,\mathbf{A}\,\mathrm{E} - \frac{1}{2}b\theta^2 \tag{5.1.3}$$

where **C**, **A**, b are material characteristics. Tensor **C,** which is a tensor of the fourth order, is called the elasticity tensor

$$\mathbf{C} = C_{\alpha\beta\gamma\delta}\,\mathbf{g}_\alpha \otimes \mathbf{g}_\beta \otimes \mathbf{g}_\gamma \otimes \mathbf{g}_\delta \tag{5.1.4}$$

Tensor **C** is symmetric

$$C_{\alpha\beta\gamma\delta} = C_{\beta\alpha\gamma\delta} = C_{\alpha\beta\delta\gamma} = C_{\beta\alpha\delta\gamma} = C_{\gamma\delta\alpha\beta} \quad (5.1.5)$$

Appearing in Eq. (5.1.3) tensor **A** is a symmetric second-order. The term – **A** θ represents the stress variation **S** – **S**° produced by the temperature variation θ in deformation when **E** = 0.
The constitutive equations are obtained by Eqs. (5.1.1) and (5.1.3) to give

$$\mathbf{S} = \mathbf{S}^{\circ} + \mathbf{C}\,\mathbf{E} - \mathbf{A}\,\theta \quad (5.1.6)$$

$$S = S_{o} + \mathbf{A}\,\mathbf{E} - b\theta \quad (5.1.7)$$

Differentiation of Eq. (5.1.7) with respect to time t gives

$$\theta_{r}^{0}\frac{dS}{dt} = \theta_{r}^{0}\,\mathbf{A}\frac{d\,\mathbf{E}}{dt} + \theta_{r}^{0}b\frac{d\theta}{dt} \quad (5.1.8)$$

Thus the thermal equation (3.5.4) reads

$$\theta_{r}^{0}\frac{dS}{dt} = R - \mathrm{Div}\,Q + \Pi^{1} \quad (5.1.9)$$

where

$$\Pi^{1} = \mathbf{S}\frac{d\,\mathbf{E}}{dt} - S\frac{d\theta_{r}}{dt} - \frac{d\Psi}{dt} \quad (5.1.10)$$

is the intrinsic volume dissipation associated with the open system dV_r. The dissipation Π_1 due to the non-dissipative character of thermo-elastic behaviour is equal to zero.

5.2 Plastic Strain

Plastic strains are typical for welding processes. In order to describe these irreversible evolutions we have to know both external variables i.e. the strain tensor **ε**, temperature θ_r and internal variables.

Any elementary system in a present state is characterized by the stress tensor **T** and the temperature θ_r. Let d**T** and $d\theta_r$ be incremental loading variations in stress and temperature, respectively, and $d\boldsymbol{\varepsilon}$ the incremental strain.

In the elementary unloading process defined by the opposite increments – $d\mathbf{T}$, – $d\theta_r$, the reversible elastic strain increments – $d\boldsymbol{\varepsilon}^e$ is observed.
The plastic strain increment $d\boldsymbol{\varepsilon}^p$ is defined by

$$d\boldsymbol{\varepsilon} = d\boldsymbol{\varepsilon}^e + d\boldsymbol{\varepsilon}^p \tag{5.2.1}$$

In the initial configuration, for which $\boldsymbol{\varepsilon} = 0$ the initial stress $\mathbf{T}^0$ and the temperature θ_r^0 are given. The plastic strain is defined as the integrals of increments $d\boldsymbol{\varepsilon}^p$

$$\boldsymbol{\varepsilon} = \boldsymbol{\varepsilon}^e + \boldsymbol{\varepsilon}^p \tag{5.2.2}$$

The plastic strain $\boldsymbol{\varepsilon}^p$ can be measurable in the unloading process from the current state. This state restores the initial stress denoted by $\mathbf{T}^0$ and the temperature θ_r^0. According to its definition and to the experiments needed for its determination, plastic variable $\boldsymbol{\varepsilon}^p$ is an internal state variable, since it is not accessible to direct observation. The increment $d\boldsymbol{\varepsilon}$ of the external variable can be measured in experiments. The increment $d\boldsymbol{\varepsilon}^e$ of the internal variable is obtained by subtracting the successive value of the external variable increment in experiments that correspond to the opposite variation of the loading. The incremental nature of plasticity is observed as the internal plastic variable $\boldsymbol{\varepsilon}^p$ measured only as integrals of increments, contrary to the external variable $\boldsymbol{\varepsilon}$, which is directly measurable.

5.3 State Equations

The infinitesimal transformation is defined by the relation

$$\|\text{Grad}\,\mathbf{u}\| << 1 \tag{5.3.1}$$

where $\mathbf{u} = \mathbf{x} - \mathbf{X}$ is the displacement vector. In infinitesimal transformation the displacement gradients are infinitesimal and we use the linearized approximation $\boldsymbol{\varepsilon}$ of the Green-Lagrange strain tensor $\mathbf{E}$

$$2\mathbf{E} \cong 2\boldsymbol{\varepsilon} = \text{grad}\,\mathbf{u} + \text{grad}\,\mathbf{u}^T \tag{5.3.2}$$

The Jacobian J of the transformation is

$$J \cong 1 + \text{tr}\,\boldsymbol{\varepsilon} \tag{5.3.3}$$

The relation combining the Cauchy stress tensor **T** and the Piola-Kirchhoff stress tensor **S** is

$$\mathbf{T} = J^{-1}\,\mathbf{F} \cdot \mathbf{S} \cdot \mathbf{F}^{T} \tag{5.3.4}$$

where $\mathbf{F} = \mathbf{1} + \text{Grad}\,\mathbf{u}$
By the infinitesimal transformation hypothesis

$$\mathbf{T} = (-\,\text{tr}\,\boldsymbol{\varepsilon})\,\mathbf{S} + \text{grad}\,\mathbf{u} \cdot \mathbf{S} + \mathbf{S}\,\text{grad}\,\mathbf{u}^{T} \tag{5.3.5}$$

The thermodynamic states of a plastic material in infinitesimal transformation are characterized by the external variables θ_r, $\boldsymbol{\varepsilon}$ and the internal variables $\boldsymbol{\varepsilon}^p$ and m_I (I = 1, …, N). The free energy is a function of the external and internal variables

$$\Psi = \Psi\,(\theta_r, \boldsymbol{\varepsilon}, \boldsymbol{\varepsilon}^p, \mathbf{m}) \tag{5.3.6}$$

The variables **m** in Eq. (5.3.6) characterize the hardening state. The state equations are expressed in the form

$$S = -\frac{\partial \Psi}{\partial \theta_r} \qquad \mathbf{T} = \frac{\partial \Psi}{\partial \boldsymbol{\varepsilon}} \tag{5.3.7}$$

The expressions (5.3.7) are based on the normality of external variables θ_r, $\boldsymbol{\varepsilon}$ with regard to the whole set of state variables. In plastic behaviour actual evolutions of these external variables varies independently from the other and must correspond to infinitesimal elastic evolutions from the present state in which $d\boldsymbol{\varepsilon}^p = 0$. The validity of state equations (5.3.7) on the border of the elasticity domain is ensured so long as free energy Ψ is continuously differentiable with respect to the external variables. The expression of free energy Ψ is presented as the second-order expansion with respect to the variables θ_r, $\boldsymbol{\varepsilon}$, $\boldsymbol{\varepsilon}^p$ close to the initial state.

$$\begin{aligned}\Psi &= \mathbf{T}^{o}\left(\boldsymbol{\varepsilon} - \boldsymbol{\varepsilon}^p\right) - S_o\theta + \frac{1}{2}\left(\boldsymbol{\varepsilon} - \boldsymbol{\varepsilon}^p\right)\mathbf{C}\left(\boldsymbol{\varepsilon} - \boldsymbol{\varepsilon}^p\right) \\ &- \theta\mathbf{A}\left(\boldsymbol{\varepsilon} - \boldsymbol{\varepsilon}^p\right) - \frac{1}{2}\frac{C_\varepsilon}{T_o}\theta^2 + U(\mathbf{m})\end{aligned} \tag{5.3.8}$$

with $\theta = \theta_r - \theta_r^0$. From Eqs. (5.3.7) and (5.3.8) the elastic state equations are

$$\mathbf{T} = \mathbf{T}^{o} + \mathbf{C}\left(\boldsymbol{\varepsilon} - \boldsymbol{\varepsilon}^p\right) - \mathbf{A}\,\theta \tag{5.3.9}$$

$$S = S_o + \mathbf{A}\left(\boldsymbol{\varepsilon} - \boldsymbol{\varepsilon}^p\right) + C_\varepsilon \theta / \theta_r^0 \tag{5.3.10}$$

In an isotropic case Eq. (5.3.9) reads

$$\mathbf{T} = \mathbf{T}^o + \left(K - 2/3\mu\right) \operatorname{tr}\left(\boldsymbol{\varepsilon} - \boldsymbol{\varepsilon}^p\right)\mathbf{1} + 2\mu\left(\boldsymbol{\varepsilon} - \boldsymbol{\varepsilon}^p\right) - 3\alpha K \theta \mathbf{1} \tag{5.3.11}$$

$$S = S_o + 3\alpha K \operatorname{tr}\left(\boldsymbol{\varepsilon} - \boldsymbol{\varepsilon}^p\right) + C_\varepsilon \theta / \theta_r^0 \tag{5.3.12}$$

Through thermo-mechanical coupling, $-\mathbf{A}\,\theta$ represents the stress variation $\mathbf{T} - \mathbf{T}^o$ produced by temperature variation θ in deformation when $\boldsymbol{\varepsilon} - \boldsymbol{\varepsilon}^p = 0$. K appearing in Eq. (5.3.11) is the bulk modulus and C_ε is the volume heat capacity per unit of initial volume, α is the thermal dilatation coefficient and μ is the Lame constant. The above state equations can be obtained for an elastic case by assuming $\boldsymbol{\varepsilon}^p = 0$. The frozen energy U (**m**) appearing in Eq. (5.3.8) is the energy due to hardening and is assumed to be independent of the external state variable $\boldsymbol{\varepsilon}$.

Assume that the function U (**m**) is independent of temperature θ_r thus, the evolution of entropy S is decoupled from the hardening evolution and the hardening latent heat is assumed to be zero. The case when the function U depends on temperature (i.e. for a non-zero hardening latent heat) is discussed in Section 5.6. Expression (5.3.8) of free energy Ψ is based on a hypothesis of separation of energies. The sum of an energy depends only on reversible variables and frozen energy.

Introduce the definition of the reduced potential in the form

$$W = \mathbf{T}^o\left(\boldsymbol{\varepsilon} - \boldsymbol{\varepsilon}^p\right) - \theta\,\mathbf{A}\left(\boldsymbol{\varepsilon} - \boldsymbol{\varepsilon}^p\right) + \frac{1}{2}\left(\boldsymbol{\varepsilon} - \boldsymbol{\varepsilon}^p\right)\mathbf{C}_o\left(\boldsymbol{\varepsilon} - \boldsymbol{\varepsilon}^p\right) \tag{5.3.13}$$

which leads to rewriting the state equations (5.3.9) and (5.3.10) in the form

$$\mathbf{T} = \frac{\partial W}{\partial\left(\boldsymbol{\varepsilon} - \boldsymbol{\varepsilon}^p\right)} = \frac{\partial W}{\partial \boldsymbol{\varepsilon}^e} \tag{5.3.14}$$

Let W*(**T**) be the Legendre-Fenchel transform of W ($\boldsymbol{\varepsilon}^e$) defined as

$$W^* = \mathbf{T}\ \boldsymbol{\varepsilon}^e - W \tag{5.3.15}$$

where **T** and $\boldsymbol{\varepsilon}^e$ are linked by relation (5.3.14). Equations (5.3.13) – (5.3.15) yield

$$\mathrm{W}^* = \frac{1}{2}\alpha_o \mathbf{A}_o \theta^2 + \left(\mathbf{T}-\mathbf{T}^o\right)\boldsymbol{\alpha}_o\theta + \frac{1}{2}\left(\mathbf{T}-\mathbf{T}^o\right)\mathbf{C}_o^{-1}\left(\mathbf{T}-\mathbf{T}^o\right) \tag{5.3.16}$$

where

$$\boldsymbol{\alpha}_o = \mathbf{C}_o^{-1}\,\mathbf{A}_o \tag{5.3.17}$$

By the properties of the Legendre-Fenchel transform

$$\boldsymbol{\varepsilon} - \boldsymbol{\varepsilon}^p = \boldsymbol{\varepsilon}^e = \frac{\partial \mathrm{W}^*}{\partial \mathbf{T}} \tag{5.3.18}$$

The expressions (5.3.16) and (5.3.17) yield

$$\boldsymbol{\varepsilon} - \boldsymbol{\varepsilon}^p = \mathbf{C}_o^{-1}\left(\mathbf{T}-\mathbf{T}^o\right) + \boldsymbol{\alpha}_o\theta \tag{5.3.19}$$

which in the isotropic case give

$$\boldsymbol{\varepsilon} - \boldsymbol{\varepsilon}^p = \frac{1+\nu_o}{\mathrm{E}_o}\left(\mathbf{T}-\mathbf{T}^o\right) - \frac{\nu_o}{\mathrm{E}_o}\operatorname{tr}\left(\mathbf{T}-\mathbf{T}^o\right)\mathbf{1} + \alpha_o\theta\,\mathbf{1} \tag{5.3.20}$$

where ν is the Poisson ratio and E is the Young modulus with the subscript o referring to the initial state.

5.4 Plasticity Criterion

In modeling of welding process it is important to assume the appropriate material characteristic of the weld and base material. The plasticity criterion, hardening effects and plastic flow rule should be assumed in such a way to describe the process as realistic as possible.

The stress **T** at any loading state characterizes any open elementary system. The loading point (**T**) in the stress space {**T**} represents the present loading state. Denote the domain of elasticity in initial state by C_D. It contains the zero loading point (**T**) = (**0**). In the elasticity domain the strain increase remains reversible or elastic, for any path of the loading point (**T**) starting from the origin of space and lying inside this domain (Fig. 5-1.).

A hardening frozen energy is absent in ideal plastic material without any hardening effect. The initial domain of elasticity for this material is not changed by the appearance of plastic strain. The elasticity domain is identical to the initial domain, and the loading point (**T**) cannot leave this domain (Fig. 5-1.). If the loading point is and remains on the boundary of the elasticity domain C_D as illustrated by the loading path **12** in Fig. 5-1. then the evolutions of plastic strain occur. Consider a loading path leaving the boundary towards the interior of domain C_D. It can be for instance the path **23** in Fig. 5-1, corresponding to a purely elastic evolution of the elementary system. It corresponds to an elastic unloading.

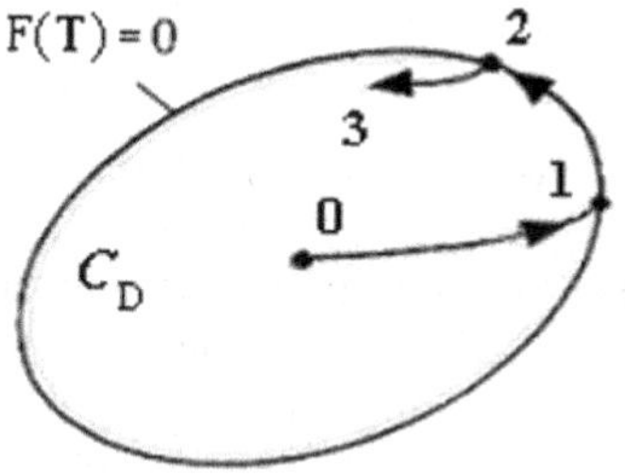

Figure 5-1. Elasticity domain of ideal plastic material

The elasticity domain is defined by a scalar function F. It is called the loading function and has **T** as its arguments. It is such that F (**T**) < 0 represents the interior of domain C_D, F (**T**) = 0 represents the boundary of domain C_D and F (**T**) > 0 represents the exterior of domain C_D. The criterion F (**T**) < 0 is the elasticity criterion. The criterion F (**T**) = 0 is the plasticity criterion. The surface in the space of loading points {**T**}, defined by F (**T**) = 0, represents the boundary of domain C_D and is called the yield locus. The plastically admissible loading state (**T**) satisfies the criterion F (**T**) ≤ 0.

The elasticity domain for hardening materials is altered by the appearance of plastic strain. In the space {**T**} of loading points, the present elasticity domain is defined as the domain arisen by the set of elastic unloading paths, or reversible loading paths, which issue from a present loading point **2**, as path **23** in Fig. 5-2.

The present loading point is not necessarily on the boundary of the present elasticity domain, such as point **3** in Fig. 5-2. There still exists an initial elasticity domain but, as soon as the loading point (**T**) reaches for the first time the boundary of the initial elasticity domain (point **1**), further loading can deform this domain while carrying it along (loading path **12**). This is the phenomenon of hardening. The present elasticity domain depends not only on the present loading point (**T**), but also on the loading path followed before, and thus on the hardening state.

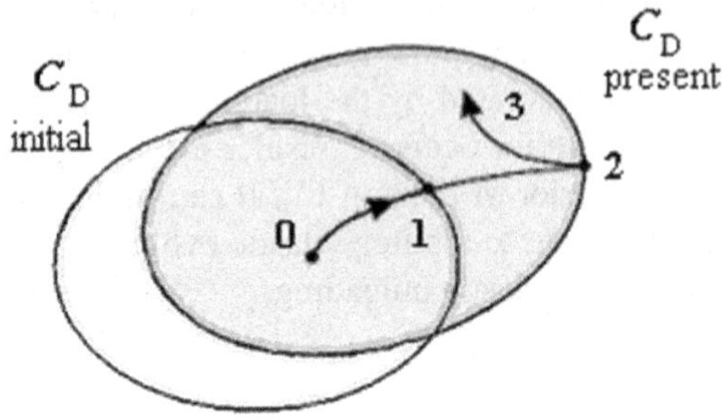

Figure 5-2. Elasticity domains of hardening material

Consider the domain of elasticity in the present state C_D. It is defined by a scalar loading function F, with arguments **T** and with some hardening parameters represented by hardening force **η**. For the hardening material, it is such that F (**T**, **η**) < 0 represents the interior of domain C_D, F (**T**, **η**) = 0 represents the boundary of domain E_D, F (**T**, **η**) > 0 represents to the exterior of domain C_D. The criterion of elasticity is expressed by F (**T**, **η**) < 0. The plasticity threshold or criterion is expressed by F (**T**, **η**) = 0. The surface defined by F (**T**, **η**) = 0, in the space of loading points {**T**}, representing the boundary of the present domain C_D is called the present yield locus.

A loading state (**T**) is the plastically admissible in the present state if it satisfies the criterion F (**T**, **η**) ≤ 0.

The present elasticity domain C_D depends on the present value of the hardening force **η**. This dependency is the basis of their experimental identification. To be useful in practice, models must involve a few hardening variables, which correspond to a few components for vector **η**. For this purpose, simple hardening models have been designed.

The first one is the isotropic hardening model. In this model the elasticity domain in space {**T**} is transformed by a homothety centred at the origin, as illustrated in Figure 5-3. The hardening force is reduced to a single scalar parameter **η** required to characterize this homothety.

The second one is the kinematic hardening model. In this model the boundaries in space {**T**} of the elasticity domain are obtained through a translation of the boundary of the initial domain. The hardening variables are the variables characterizing this translation. They reduce to a tensor parameter **η** relative to the translation with respect to the stress tensor (Fig. 5-3). The two previous hardening models can also be combined to yield an isotropic and kinematic hardening model, as illustrated in Figure 5-3. As defined in this section, the hardening force **η** represents only a set of variables well suited for mathematical description of the observed evolution of the elasticity domain, and thus may not yet be considered as a thermodynamic force.

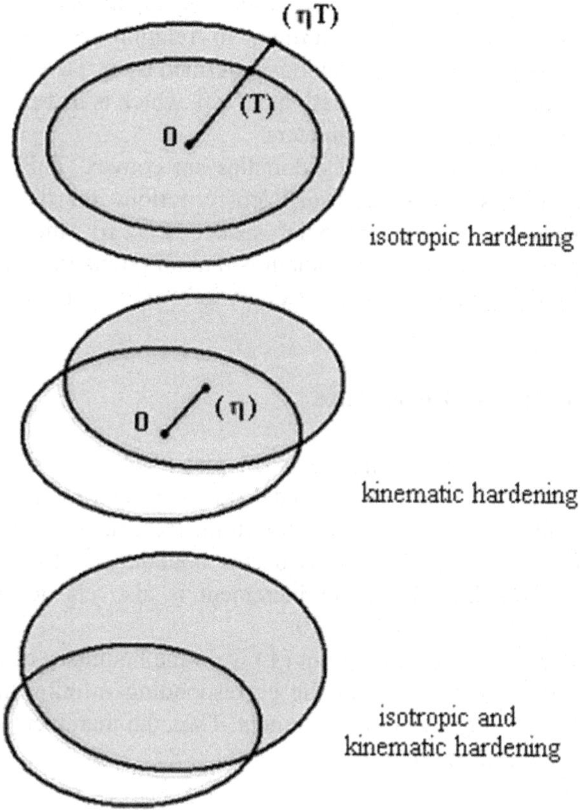

Figure 5-3. Usual hardening models

The hardening force $\boldsymbol{\eta}$ can be associated with a set of hardening variables $\mathbf{m}$ by relations

$$\boldsymbol{\eta} = -\frac{\partial U}{\partial \mathbf{m}} \tag{5.4.1}$$

The frozen energy U, is identified with the help of independent calorimetric measurements. The absolute temperature θ_r should not appear explicitly in the mathematical description of the elasticity domain. Its influence on hardening phenomena can be considered through a dependence of η upon θ_r.

According to relation (5.4.1) the temperature θ_r can be an argument of energy U. The influence of T on hardening phenomena will not be considered in the following. According to relation (5.4.1), the virgin hardening state, which has been previously defined by $\eta = 0$ as illustrated in Fig. 5-3, now corresponds to the relation dU = 0, which is irrespective of any particular choice of hardening parameters.

The initial and present elasticity domains are convex. This property of convexity constitutes one of the sufficient criterions for the stability of plastic materials. In the loading point space $\{\mathbf{T} \times \eta\}$, the fundamental geometrical property of a convex domain is that all points of a segment of a line that joins two points on the boundary of the domain lie inside this domain.

5.5 The Plastic Flow Rule

The plasticity criterion indicates when plastic phenomena occur. The plastic flow rule indicates how. If the loading point (**T**) lies within the elasticity domain C_D, F (**T**) < 0, then the strain increments d**ε** are elastic or reversible. If the loading point (**T**) is on the boundary of C_D but leaves it during elastic unloading the strain increment is also elastic. The elastic unloading criterion is F = 0 and dF < 0.

In the case when the loading point (**T**) is on the boundary of the domain C_D, i.e. F = dF = 0 we say that the corresponding infinitesimal loading increment d**T** is a neutral loading increment. Then the increment d**ε** may not be elastic. This is expressed by

$$d\boldsymbol{\varepsilon}^p = 0 \quad \text{if} \quad F(\mathbf{T}) < 0 \quad \text{or if} \quad F(\mathbf{T}) = 0 \quad dF = \frac{\partial F}{\partial \mathbf{T}} d\mathbf{T} < 0$$

$$d\boldsymbol{\varepsilon} = d\boldsymbol{\varepsilon}^e + d\boldsymbol{\varepsilon}^p \quad \text{if} \quad F = dF = 0 \tag{5.5.1}$$

The notation dF states that the function F is differentiable with respect to its arguments, at the considered loading point (**T**) on the boundary of the domain C_D.

Let ∂F be the subdifferential of F with respect to (**T**) at the present loading point (**T**). The subdifferential ∂F at a regular point on the boundary of domain C_D defined by F (**T**) = 0 reduces to the gradient of F with respect to (**T**) and thus corresponds geometrically at this point to the outward normal $\partial F/\partial \mathbf{T}$ to the boundary. At a singular point in the boundary of domain C_D a convex loading function shows that the set $\alpha\partial F$ with $\alpha \geq 0$ is constituted by the cone F outward normals to the boundary of domain C_D.

The expression (5.5.1) indicates when an evolution of the plastic strain occurs. All values for $d\boldsymbol{\varepsilon}^p$ are not admissible since they must ensure the non-negativeness of intrinsic dissipation Π_1. Since the intrinsic dissipation reduces to the plastic work rate for ideal plastic materials, the non-negativeness of intrinsic dissipation, or equivalently of the infinitesimal plastic work rate requires the relation for the plastic increments $d\boldsymbol{\varepsilon}^p$

$$\mathbf{T}\, d\boldsymbol{\varepsilon}^p \geq 0 \tag{5.5.2}$$

Consider the relation

$$d\boldsymbol{\varepsilon}^p \in d\lambda\ \mathcal{G}(\mathbf{T}) \qquad d\lambda \geq 0 \tag{5.5.3}$$

where $d\lambda$ is the so-called plastic multiplier. The plastic multiplier is a non-negative scalar factor. The set of thermodynamically admissible directions in the loading point space $\{\mathbf{T}\}$ for the vector $(d\boldsymbol{\varepsilon}^p)$ ensuring the non-negativeness of its scalar product $\mathbf{T}\, d\boldsymbol{\varepsilon}^p$ is represented by the set $\mathcal{G}(\mathbf{T})$.
Fig. 5-4. illustrates the set $\mathcal{G}(\mathbf{T})$ in the loading point space $\{\mathbf{T}\}$. The admissible directions for the vector $(d\boldsymbol{\varepsilon}^p)$ are independent of the plastic criterion.

The plastic flow rule derived from the non-negativeness of intrinsic dissipation for an ideal plastic material can be summarized by

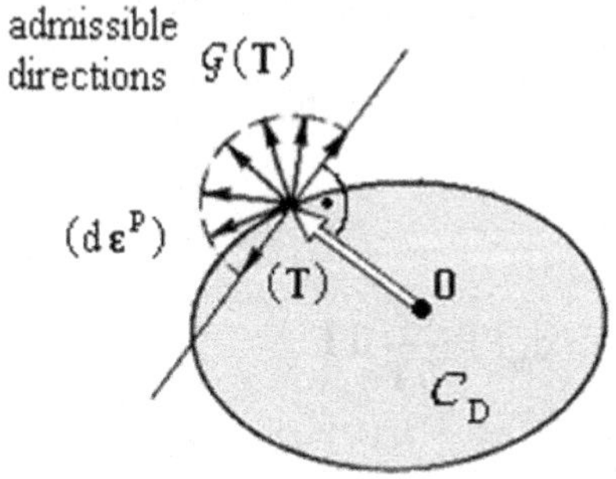

Figure 5-4. Admissible directions of plastic increments for ideal plastic material

$$d\boldsymbol{\varepsilon}^p \in d\lambda\ \mathcal{G}(\mathbf{T}) \quad d\lambda \geq 0 \quad \text{where} \quad \begin{array}{l} d\lambda \geq 0 \text{ if } F = 0 \text{ and } dF = 0 \\ d\lambda = 0 \text{ if } F < 0 \text{ or } dF < 0 \end{array} \tag{5.5.4}$$

If the plastic increments $d\boldsymbol{\varepsilon}^p$ have non-zero values, then we say that the loading is plastic.

The evolution for hardening materials is elastic with no change of the hardening state, if in the space $\{\mathbf{T} \times \boldsymbol{\eta}\}$ the present extended loading point $(\mathbf{T}, \boldsymbol{\eta})$ lies inside the fixed domain C ($F < 0$) or on its boundary $F = 0$ and leaves it ($dF < 0$). The above can be written in the form

$$(d\boldsymbol{\varepsilon}^p, d\mathbf{m}) \in d\lambda\ \mathcal{G}(\mathbf{T}, \boldsymbol{\eta}) \quad d\lambda \geq 0 \quad \text{where} \quad \begin{array}{l} d\lambda \geq 0 \text{ if } F = 0 \text{ and } dF = 0 \\ d\lambda = 0 \text{ if } F = 0 \text{ or } dF = 0 \end{array} \tag{5.5.5}$$

If the increments $(d\boldsymbol{\varepsilon}^p, d\mathbf{m})$ have non-zero values, then we say that the loading is plastic. The scalar $d\lambda$ in Eq. (5.5.5) is the plastic multiplier, and $\mathcal{G}(\mathbf{T}, \boldsymbol{\eta})$ are the set of thermodynamically admissible directions for the increments $(d\boldsymbol{\varepsilon}^p, d\mathbf{m})$ satisfying the relation of the non-negativeness of the intrinsic dissipation

$$\mathbf{T}\, d\boldsymbol{\varepsilon}^p + \boldsymbol{\eta} \cdot d\mathbf{m} \geq 0 \tag{5.5.6}$$

The increment $d\mathbf{m}$ of the hardening variable is written in the form

$$d\mathbf{m} = d\lambda\ \mathcal{G}_{\mathbf{m}}(\mathbf{T}, \boldsymbol{\eta}) \tag{5.5.7}$$

In the above $\mathcal{G}_{\mathbf{m}}(\mathbf{T}, \boldsymbol{\eta})$ are the actual directions of the permissible directions for $d\mathbf{m}$.

In the case of hardening plasticity the loading function depends on the hardening force $\boldsymbol{\eta}$. Then the expression for dF is

$$dF = d_{\eta}F + d_{T}F \tag{5.5.8}$$

with

$$d_{\eta}F = \frac{\partial F}{\partial \mathbf{T}}\, d\mathbf{T} \tag{5.5.9}$$

$$d_{T}F = \frac{\partial F}{\partial \boldsymbol{\eta}}\, d\boldsymbol{\eta} \tag{5.5.10}$$

If the plastic increments $d\boldsymbol{\varepsilon}^p$ have non-zero values, then we say that the loading is plastic.

In the above $d_\eta F$ is the differential of function F with the hardening force η being held constant. The expression (5.5.10) can be written as $d_T F = -H d\lambda$, where H is the hardening modulus defined by

$$H = -\frac{\partial F}{\partial \eta} \cdot \frac{\partial \eta}{\partial \mathbf{m}} \cdot \frac{\partial \mathbf{m}}{\partial \lambda} = \frac{\partial F}{\partial \eta} \cdot \frac{\partial^2 U}{\partial \mathbf{m}^2} \cdot \mathcal{G}_{\mathbf{m}}(\mathbf{T}, \eta) \tag{5.5.11}$$

The definition (5.5.10) of $d_T F$ and the relation (5.5.11) show that H has a stress unit. In order to specify the plastic flow rule, we need to examine the specific case of hardening ($H > 0$) and the case of softening ($H < 0$).
In the case of hardening

$$\left(d\boldsymbol{\varepsilon}^p, d\mathbf{m}\right) \in \frac{\partial_\eta F}{H} \mathcal{G}(\mathbf{T}, \eta) \quad \text{if } F = 0 \quad \text{and } d_\eta F > 0$$

$$d\boldsymbol{\varepsilon}^p = d\mathbf{m} = 0 \quad \text{if } F < 0 \;\; \text{or if } F = 0 \;\; \text{and } d_\eta F \leq 0 \tag{5.5.12}$$

In the case of softening

$$\left(d\boldsymbol{\varepsilon}^p, d\mathbf{m}\right) \in \frac{\partial_\eta F}{H} \mathcal{G}(\mathbf{T}, \eta) \text{ or } d\boldsymbol{\varepsilon}^p = d\mathbf{m} = 0 \quad \text{if } F = 0 \quad \text{and } d_\eta F < 0$$

$$d\boldsymbol{\varepsilon}^p = d\mathbf{m} = 0 \quad \text{if } F < 0 \quad \text{or if } F = d_\eta F = 0 \tag{5.5.13}$$

The relations (5.5.12) and (5.5.13) for hardening or softening materials show that if the loading point (**T**) is and remains on the boundary of the present elasticity domain C_D (i.e. if $F = d_\eta F = 0$), the evolution is purely elastic without a change in the hardening state (i.e. $d\boldsymbol{\varepsilon}^p = d\mathbf{m} = d\eta = 0$). Then we state that the loading increment d**T** is neutral. This situation is illustrated in Fig. 5.5.

The interpretation of the hardening sign can be explained as follows. In the case of hardening, expression (5.5.12) states that $d_\eta F$ is strictly positive for a plastic evolution to occur. The geometrical meaning of $d_\eta F$ (the scalar product of the present loading increment (d**T**) with outward unit normal ($\partial F/\partial \mathbf{T}$)) shows that the vector of present loading increment d**T** in the space {**T**} must be oriented outwards with regard to the present elasticity domain C_D at the present loading point (d**T**). In fact, in the case of a regular point, its scalar product with unit outward normal **n** must be positive. In other words, the new loading point (**T** + d**T**) escapes from the present elasticity domain

C_D, while carrying it along. Fig. 5-5 illustrates this geometrical interpretation in the space {**T**} and Fig. 5-6 in the space {**T** × **η**}.

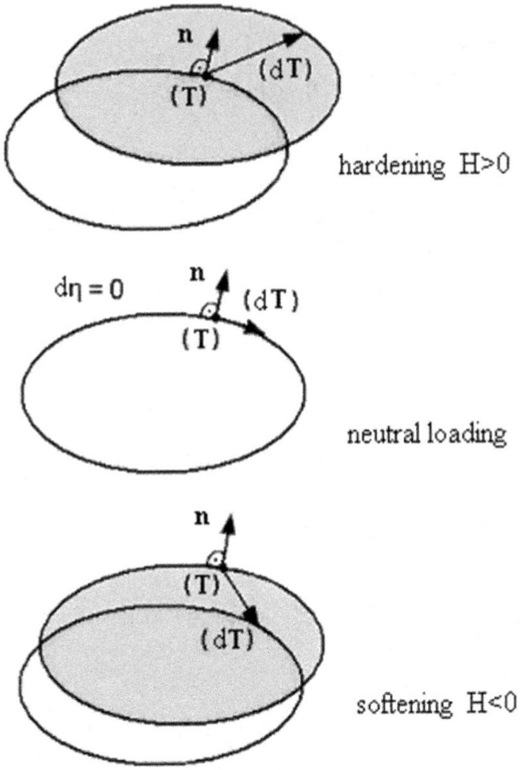

Figure 5-5. Interpretation of hardening in the space {**T**}

In the case of softening, expression (5.5.13) states that $d_\eta F$ is strictly negative for a plastic evolution to occur. Consider the loading point (**T** + d**T**). This loading point carries the elasticity domain inside C_D when softening occurs. Taking into account the hardening sign, we state that softening is negative hardening.

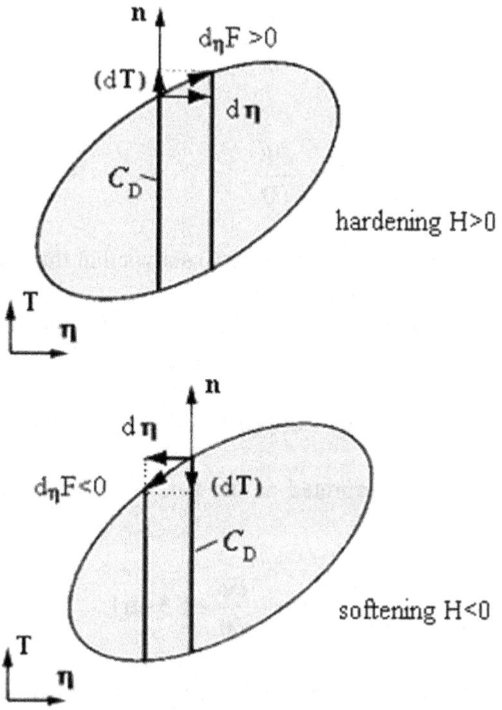

Figure 5-6. Hardening sign in the space $\{\mathbf{T} \times \boldsymbol{\eta}\}$

5.6 Thermal Hardening

Consider the case where the hardening force $\boldsymbol{\eta}$ representing the evolution of the elasticity domain is independent of the temperature. The expression of free energy Ψ in this case is of the form

$$\Psi = \Psi\,(\mathrm{T}, \boldsymbol{\varepsilon}, \boldsymbol{\varepsilon}^{\mathrm{p}}, \mathbf{m}) = \varphi\,(\boldsymbol{\varepsilon} - \boldsymbol{\varepsilon}^{\mathrm{p}}) + \mathrm{U}\,(\mathbf{m}) \tag{5.6.1}$$

and

$$\boldsymbol{\eta} = -\frac{\partial \Psi}{\partial \mathbf{m}} \qquad \mathrm{S} = -\frac{\partial \Psi}{\partial \theta_{\mathrm{r}}} \tag{5.6.2}$$

In the case of thermal hardening effects, in the expression of free energy Ψ the temperature $\theta = \theta_{\mathrm{r}} - \theta_{\mathrm{r}}^{0}$ is included

$$\Psi = \Psi\,(\theta = \theta_r - \theta_r^0,\ \boldsymbol{\varepsilon},\ \boldsymbol{\varepsilon}^p,\ \mathbf{m}) = \varphi\,(\boldsymbol{\varepsilon} - \boldsymbol{\varepsilon}^p,\ \theta\,) + U^*\,(\mathbf{m},\ \theta) \tag{5.6.3}$$

and

$$\eta = -\frac{\partial \Psi}{\partial \mathbf{m}} \qquad S = -\frac{\partial \Psi}{\partial \theta} \tag{5.6.4}$$

Substituting expression (5.6.3) into (5.6.4) shows that the hardening force η depends on temperature variation $\theta = \theta_r - \theta_r^0$.

Assume that the temperature variation is small. The function U^* ($\mathbf{m}$, θ) can be expressed as

$$U^*(\mathbf{m},\ \theta) = U(\mathbf{m}) - \theta S^*(\mathbf{m}) \tag{5.6.5}$$

The function U($\mathbf{m}$) is interpreted as the frozen free energy. By (5.6.3), (5.6.4) and (5.6.5) we get

$$\eta = -\frac{\partial U}{\partial \mathbf{m}} + \theta \frac{\partial S^*}{\partial \mathbf{m}} \qquad S = -\frac{\partial \varphi}{\partial \theta} + S^*(\mathbf{m}) \tag{5.6.6}$$

Consider the elasticity domain C_D in the loading space {**T**} defined by

$$F = F\,(\mathbf{T},\ \eta) \le 0 \tag{5.6.7}$$

According to Eq. (5.6.6), due to the temperature variation dθ, the hardening force η and the elasticity domain C_D may change. Thermal hardening occurs for a zero frozen free energy U ($\mathbf{m}$) = 0. An inverse temperature variation – dθ restores the previous elasticity domain. The second relation (5.6.6) shows also that there is an unrecovered change in entropy, yielding the frozen entropy term $S^*(\mathbf{m})$. With respect to the linearized thermal equation (3.8.6) the quality $\theta_r^0\,S^*(\mathbf{m})$ can easily be related to a hardening latent heat effect. The flow rule can be expressed as

$$(d\boldsymbol{\varepsilon}^p,\ d\mathbf{m}) \in d\lambda\ \mathcal{G}(\mathbf{T},\ \eta)\ d\lambda \ge 0 \ \text{ where } \begin{array}{lll} d\lambda \ge 0 & \text{if } F = 0 & \text{and } dF = 0 \\ d\lambda = 0 & \text{if } F < 0 & \text{or } dF < 0 \end{array} \tag{5.6.8}$$

where the plastic multiplier dλ and the hardening modulus H are now expressed in the form

$$d\lambda = \frac{d_{\mathbf{m}} F}{H} \qquad H = -\frac{\partial F}{\partial \eta} \cdot \frac{\partial \eta}{\partial \mathbf{m}} \cdot \frac{\partial \mathbf{m}}{\partial \lambda} = \frac{\partial F}{\partial \eta} \cdot \frac{\partial^2 U}{\partial \mathbf{m}^2} \cdot \mathcal{G}_{\mathbf{m}}(\mathbf{T}, \eta) \tag{5.6.9}$$

Substituting (5.6.6) into (5.6.7) $d_\eta F$ can be expressed in the form

$$d_{\mathbf{m}} F = \frac{\partial F}{\partial \mathbf{T}} d\mathbf{T} + \frac{\partial F}{\partial \eta} \cdot \frac{dS^*}{\partial \mathbf{m}} d\theta \tag{5.6.10}$$

In order to express the loading function F in terms of θ and **m** we use Eq. (5.6.6) and Eq. (5.6.7). Assume dF = 0, then Eqs. (5.6.8) and (5.6.9) give

$$F(\mathbf{T}, \theta, \mathbf{m} + d\mathbf{m}) = F(\mathbf{T}, \theta, \mathbf{m}) + \frac{\partial F}{\partial \mathbf{m}} \cdot d\mathbf{m} = F(\mathbf{T}, \theta, \mathbf{m}) - H\, d\lambda \tag{5.6.11}$$

5.7 The Hypothesis of Maximal Plastic Work

The admissible directions of the plastic strain increment $d\boldsymbol{\varepsilon}^p$ are defined by the hypothesis of maximal plastic work. This hypothesis is formulated as follows.

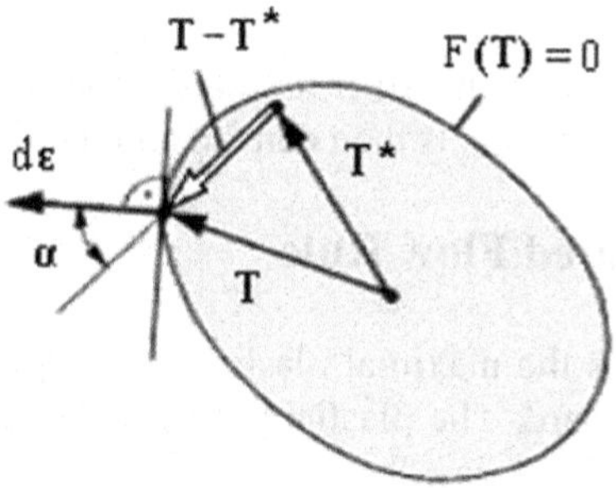

Figure 5-7. Geometrical interpretation of the hypothesis of maximal plastic work

Let (**T**) be the present loading state in an elementary material system being non-exterior to the present elasticity domain i.e. $(\mathbf{T}) \in C_D$ and $d\boldsymbol{\varepsilon}^p$ be the associated plastic increment of strain. Let (**T***) be another loading state being non-exterior to the present elasticity domain i.e. $(\mathbf{T}^*) \in C_D$. The hypothesis of maximal plastic work is expressed by

$$(\mathbf{T} - \mathbf{T}^*)\, d\boldsymbol{\varepsilon}^p \geq 0 \tag{5.7.1}$$

for every $(\mathbf{T}^*) \in C_D$.

This means that the loading state (**T**) to be associated with the present plastic increment of strain ($d\boldsymbol{\varepsilon}^p$) is the one among those admitted by the present plastic criterion, maximizing the infinitesimal plastic work.

Fig 5-7 represents the geometrical interpretation of the hypothesis of maximal plastic work. The directions of vectors ($\mathbf{T} - \mathbf{T}^*$) and $d\boldsymbol{\varepsilon}^p$ form the non-obtuse angle.

The hypothesis of maximal plastic work can be geometrically represented in the loading point space {**T**} by the convexity of the elasticity domain E_D and by the directions permissible for the plastic increment ($d\boldsymbol{\varepsilon}^p$).

The flow rule reads

$$d\boldsymbol{\varepsilon}^p \in d\lambda\, \partial F(\mathbf{T}) \qquad \text{or } d\boldsymbol{\varepsilon}^p \in d\lambda\, \partial_\eta F(\mathbf{T}, \boldsymbol{\eta}) \qquad d\lambda \geq 0 \tag{5.7.2}$$

where $\partial F(\mathbf{T})$ and $\partial_\eta F(\mathbf{T}, \boldsymbol{\eta})$ represent the subdifferential of $F(\mathbf{T})$ and $\partial_\eta F(\mathbf{T}, \boldsymbol{\eta})$ respectively with respect to **T**.

The hypothesis of maximal plastic work implies both the normality of the flow rule and the convexity of the plastic criterion. The hypothesis of maximal plastic work and the hypothesis of both the normality of the flow rule and the convexity of the plastic criterion with respect to **T** are equivalent.

A material is said to be standard if it satisfies the maximal plastic work hypothesis. The plastic criterion is then convex and the flow rule is normal.

5.8 The Associated Flow Rule

If a material satisfies the maximal plastic work hypothesis, then we say that the material is standard. The plasticity criterion is then convex and the flow rule is normal and we say that the flow rule is associated with the criterion of plasticity.

Consider an ideal plastic standard material. Eqs. (5.5.4) and (5.7.2) lead to write the flow rule as

$$d\boldsymbol{\varepsilon}^p \in d\lambda\, \partial F(\mathbf{T}) \quad d\lambda \geq 0 \text{ where } \begin{array}{lll} d\lambda \geq 0 & \text{if } F = 0 & \text{and } dF = 0 \\ d\lambda = 0 & \text{if } F < 0 & \text{and } dF < 0 \end{array} \tag{5.8.1}$$

and for any regular point on the boundary of the elasticity domain in the case of plastic loading

$$d\boldsymbol{\varepsilon}^p = d\lambda \frac{\partial F}{\partial \mathbf{T}} \quad d\lambda \geq 0 \quad \text{if } F = dF = 0 \tag{5.8.2}$$

Consider a hardening standard material. Eqs. (5.5.12), (5.5.13) and (5.8.2) give for a hardening standard material

$$(d\boldsymbol{\varepsilon}^p, d\mathbf{m}) \in \frac{d_\eta F}{H}\partial_\eta F(\mathbf{T},\boldsymbol{\eta}) \quad \text{if } F = 0 \text{ and } d_\eta F > 0$$

$$d\boldsymbol{\varepsilon}^p = d\mathbf{m} = 0 \quad \text{if } F < 0 \quad \text{or if } F = 0 \text{ and } d_\eta F \le 0 \tag{5.8.3}$$

For softening ($H < 0$) we get

$$(d\boldsymbol{\varepsilon}^p, d\mathbf{m}) \in \frac{d_\eta F(\mathbf{T},\boldsymbol{\eta})}{H}\partial_\eta F(\mathbf{T},\boldsymbol{\eta}) \text{ or } d\boldsymbol{\varepsilon}^p = d\mathbf{m} = 0 \quad \text{if } F = 0 \text{ and } d_\eta F < 0$$

$$d\boldsymbol{\varepsilon}^p = d\mathbf{m} = 0 \quad \text{if } F < 0 \quad \text{or if } F = 0 \text{ and } d_\eta F = 0 \tag{5.8.4}$$

Consider a regular point on the boundary of the elasticity domain and non-zero plastic increments $d\boldsymbol{\varepsilon}^p$. Then we have

$$d\boldsymbol{\varepsilon}^p = \frac{d_\eta F}{H}\left(\frac{\partial F}{\partial \mathbf{T}}\right) \tag{5.8.5}$$

5.9 Incremental Formulation for Thermal Hardening

In solution of welding thermomechanical problems an important is the incremental formulation for thermal hardening. Consider Eq. (5.3.9) with its incremental form

$$d\mathbf{T} = \mathbf{C}\,(d\boldsymbol{\varepsilon} - d\boldsymbol{\varepsilon}^p) - \mathbf{A}d\theta \tag{5.9.1}$$

By Eq. (5.5.12) and (5.6.10) we get

$$d\mathbf{T} = \mathbf{C}\,d\boldsymbol{\varepsilon} - \mathbf{A}d\theta - \frac{1}{H}\left(\frac{\partial F}{\partial \mathbf{T}} + \frac{\partial F}{\partial \boldsymbol{\eta}}\frac{\partial S^*}{\partial \mathbf{m}}d\theta\right)\mathbf{C}\frac{\partial F}{\partial \mathbf{T}} \tag{5.9.2}$$

By contracting Eq. (5.9.1) with the tensor $\partial F/\partial \mathbf{T}$ and adding the result to the quantity $\dfrac{\partial F}{\partial \boldsymbol{\eta}}\dfrac{\partial S^*}{\partial \mathbf{m}}d\theta$ gives

$$\frac{\partial F}{\partial \mathbf{T}}+\frac{\partial F}{\partial \eta}\frac{\partial S^{f}}{\partial \mathbf{m}}d\theta=\frac{\dfrac{\partial F}{\partial \mathbf{T}}\mathbf{C}\,d\boldsymbol{\varepsilon}+\left(\dfrac{\partial F}{\partial \eta}\dfrac{\partial S^{*}}{\partial \mathbf{m}}-\dfrac{\partial F}{\partial \mathbf{T}}\mathbf{A}\right)d\theta}{1+\dfrac{1}{H}\dfrac{\partial F}{\partial \mathbf{T}}\mathbf{C}\dfrac{\partial F}{\partial \mathbf{T}}} \tag{5.9.3}$$

By substituting (5.9.3) into (5.9.2) we get

$$d\mathbf{T}=\mathbf{C}\,d\boldsymbol{\varepsilon}-\mathbf{A}\,d\theta-Y(F)\frac{\left[\dfrac{\partial F}{\partial \mathbf{T}}\mathbf{C}\,d\boldsymbol{\varepsilon}+\left(\dfrac{\partial F}{\partial \eta}\dfrac{\partial S^{*}}{\partial \mathbf{m}}-\mathbf{A}\dfrac{\partial F}{\partial \mathbf{T}}\right)d\theta\right]\mathbf{C}\dfrac{\partial F}{\partial \mathbf{T}}}{H+\dfrac{\partial F}{\partial \mathbf{T}}\mathbf{C}\dfrac{\partial F}{\partial \mathbf{T}}} \tag{5.9.4}$$

where Y(x) is the Heaviside function defined by

$$Y(x)=0 \quad \text{if} \quad x<0 \quad \text{and} \quad Y(x)=1 \quad \text{if} \quad x\geq 0 \tag{5.9.5}$$

By (5.9.4) we get

$$d\mathbf{T}=\frac{\partial W^{*}}{\partial (d\boldsymbol{\varepsilon})} \tag{5.9.6}$$

where W*(d**ε**) is the incremental potential defined by

$$\begin{aligned} W^{*}=&\frac{1}{2}d\boldsymbol{\varepsilon}\,\mathbf{C}\,d\boldsymbol{\varepsilon}-d\theta\,\mathbf{A}\,d\boldsymbol{\varepsilon}\\ &-\frac{1}{2}Y(F)\frac{\left[\dfrac{\partial F}{\partial \mathbf{T}}\mathbf{C}\,d\boldsymbol{\varepsilon}+\left(\dfrac{\partial F}{\partial \eta}\dfrac{\partial S^{*}}{\partial \mathbf{m}}-\mathbf{A}\dfrac{\partial F}{\partial \mathbf{T}}\right)d\theta\right]^{2}}{H+\dfrac{\partial F}{\partial \mathbf{T}}\mathbf{C}\dfrac{\partial F}{\partial \mathbf{T}}} \end{aligned} \tag{5.9.7}$$

5.10 Models of Plasticity

In modeling of welding processes we should consider the appropriate models of plasticity. In isotropic material the loading function F involves the principal components of symmetric stress tensor **T**, i.e. the three principal stresses T_1, T_2 and T_3. The principal stresses can be expressed in terms of the three first invariants of the stress tensor. We denote the first invariant of the stress tensor by J_{1T}, the second invariant of the stress deviator tensor $\mathbf{T}' = \mathbf{T} - (\mathrm{tr}\,\mathbf{T}/3)\mathbf{1}$ by $J_{2T'}$ and the third invariant of the stress deviator tensor by $J_{3T'}$

$$J_{1T}(\mathbf{T}) = \mathrm{tr}\,\mathbf{T} = 3p \tag{5.10.1}$$

$$h^2 = J_{2\mathbf{T}'}(\mathbf{T}) = \frac{1}{2} T'_{ij}\, T'_{ji} = \frac{1}{2}\mathrm{tr}(\mathbf{T}' \cdot \mathbf{T}') \tag{5.10.2}$$

$$J_{3\mathbf{T}'}(\mathbf{T}) = \frac{1}{3} T'_{ij}\, T'_{jk}\, T'_{kl} \tag{5.10.3}$$

By (5.10.2) we have

$$h^2 = \frac{1}{6}\left[(T_1 - T_2)^2 + (T_1 - T_3)^2 + (T_2 - T_3)^2\right] \tag{5.10.4}$$

The loading function for isotropic models of plasticity can be represented by

$$F = F(h, p) \tag{5.10.5}$$

In the case of isotropic hardening materials, the loading function F is expressed in the form

$$F = F(h, p, \boldsymbol{\eta}) \tag{5.10.6}$$

where $\boldsymbol{\eta}$ is the hardening force describing the evolution of the yield surface in loading point space $\{T_i\}$. In isotropic hardening the yield surface is derived through a homothety of center $\mathbf{0}$ in the loading point space $\{T_i\}$. Then the hardening force $\boldsymbol{\eta}$ reduces to a scalar variable $\boldsymbol{\eta}$ which defines this homothety.

The loading function given by (5.10.6) can be expressed as a homogeneous polynome of degree n with regard to h and η

$$F = F(h, p, \eta) = \eta^n F(h/\eta, p/\eta, 1) \tag{5.10.7}$$

where by convection η is specified as the ratio of the homothety that transforms the yield surface defined by $\eta = 1$ into the present yield surface. In kinematic hardening, the yield surfaces are defined from each other through a translation in the loading point space $\{T_i\}$. The hardening force $\boldsymbol{\eta}$ reduces to a second-order symmetric tensor $\boldsymbol{\eta}$ that defines this translation

$$F = F\,[J_{2T'}(\mathbf{T} + \boldsymbol{\eta}),\, J_{1T}(\mathbf{T} + \boldsymbol{\eta})] \tag{5.10.8}$$

In space $\{T_i\}$ vector $(\boldsymbol{\eta})$ represents the vector of translation that transforms the yield surface defined by $(\boldsymbol{\eta}) = (0)$ into the present yield surface.

Assume the convex loading function for the isotropic plastic material

$$F\,(h, p) = h + \xi\, p - q \tag{5.10.9}$$

where ξ and q are material characteristics. The constant q is necessarily non-negative to ensure that the zero loading point satisfies $F\,(0,0) \le 0$. The coefficient ξ is non-negative to describe an infinite tensile stress. The yield surface given by (5.10.9) is an axisymmetric surface around the trisector in principal stress space $\{T_i\}$. If $\xi = 0$ the loading function reduces to the Huber-Mises loading function.

The form of the Huber-Mises loading function is

$$F = \frac{1}{\sqrt{3}}\sqrt{T_{11}^2 + T_{22}^2 + T_{33}^2 - T_{11}T_{22} - T_{22} - T_{33} - T_{33}T_{11} + 3\left(T_{12}^2 + T_{23}^2 + T_{31}^2\right)} - q \tag{5.10.10}$$

or for principal directions

$$F = \frac{1}{\sqrt{3}}\sqrt{T_1^2 + T_2^2 + T_3^2 - T_1T_2 - T_2T_3 - T_3T_1} - q \tag{5.10.11}$$

The Huber-Mises loading function can be transformed to the equivalent forms if we introduce material parameter $q = \frac{1}{\sqrt{3}} T_0$, where T_0 is the yield point of the material in uniaxial tension.

Then the Huber-Mises loading function is expressed in the frequently met form

$$F = \frac{1}{\sqrt{6}}\sqrt{\left(T_{11} - T_{22}\right)^2 + \left(T_{22} - T_{33}\right)^2 + \left(T_{33} - T_{11}\right)^2 + 6\left(T_{12}^2 + T_{23}^2 + T_{31}^2\right)} - \frac{1}{\sqrt{3}} T_0 \tag{5.10.12}$$

or

$$F = \frac{1}{\sqrt{6}}\sqrt{(T_1 - T_2)^2 + (T_2 - T_3)^2 + (T_3 - T_1)^2} - \frac{1}{\sqrt{3}}T_0 \qquad (5.10.13)$$

The previous loading function (5.10.9) for an isotropic material that exhibits isotropic hardening is extended in the form

$$F = h + \xi p - \eta q \qquad (5.10.14)$$

The yield surface that corresponds to the loading function (5.10.14) is an axisymmetric cone around the trisector in principal stress space $\{T_i\}$. In this space, the three coordinates of the vertex of the cone are equal to $q\eta/\xi$. In the loading point space $\{T_i\}$ the singular points are located on a line given by the equation $T_1 = T_2 = T_3 = q\eta/\xi$. The present yield surface is the transform of the surface defined by $\eta = 1$, through the homothety centred at the origin with η as the ratio. For an isotropic material exhibiting kinematic hardening

$$F = \sqrt{\frac{1}{2}(s_{ij} + \beta_{ij})(s_{ji} + \beta_{ji})} + \xi(p + \eta_{ii}) - q \qquad (5.10.15)$$

where the symmetric tensor $\boldsymbol{\beta}$ is the deviator of the tensor

$$\boldsymbol{\beta} = \boldsymbol{\eta} - 1/3\,(\mathrm{tr}\,\boldsymbol{\eta})\mathbf{1} \qquad (5.10.16)$$

In the loading point space $\{T_i\}$ the present yield surface is the transform of the surface defined by $\boldsymbol{\eta} = 0$ through the translation defined by the vector with $\boldsymbol{\eta}$ and the eigenvalues η_1, η_2, η_3 of the tensor $\boldsymbol{\eta}$ as components. In the loading point space $\{T_i\}$ the singular points are located on the line given by the equation $T_1 + \eta_1 = T_2 + \eta_2 = T_3 + \eta_3 = q/\xi$. The expression (5.10.14) and (5.10.15) may be combined to obtain

$$F = \sqrt{\frac{1}{2}(s_{ij} + \beta_{ij})(s_{ji} + \beta_{ji})} + \xi(p + \eta_{ii}) - q\eta' \qquad (5.10.17)$$

In the loading space $\{T_i\}$ the present yield surface is the transform of the surface defined by $\boldsymbol{\eta} = 0$ and $\eta' = 1$ through the translation defined by the vector with (η_1, η_2, η_3) as components followed by the homothety of ratio η' and centred at the point (η_1, η_2, η_3), the translated point of origin.

In the loading point space $\{T_i\}$ the singular points are located on the line given by the equation

$$T_1 + \eta_1 = T_2 + \eta_2 = T_3 + \eta_3 = q\eta'/\xi \qquad (5.10.18)$$

Part II

NUMERICAL ANALYSIS OF WELDING PROBLEMS

Chapter 6

NUMERICAL METHODS IN THERMOMECHANICS

6.1 Introduction

Most of the thermo-mechanical problems in welding can only be solved by using numerical procedures. In realistic model the thermal conductivity and specific heat should be considered as a function of temperature which complicates analytical solutions. Because of possible phase change the analytical methods have a limited range of applications. In most types of welding melting will occur and there will also be convective heat transfer in addition to conductive heat transfer. The heat sources in welding in realistic model are not concentrated in point or line.

In this section the general data concerning the numerical solving methods applied to welding processes are given. The non-linearity of the constitutive equations describing welding process and its time-dependence require the advanced solution technique.

6.2 Finite-Element Solution of Heat Flow Equations

6.2.1 Weighted Residual Method

Heat transfer problems can be formulated either by a given differential equation with boundary conditions or by a given functional equivalent to the differential equation. The weighted residual method is an approximate solution method for differential equations. This method widens the range of problems amenable to solution since it does not require a variational formulation of the problem. Consider the heat flow equation

$$\nabla(k\,\nabla\theta)+r=C_{\Delta}\frac{\partial\theta}{\partial t} \tag{6.2.1}$$

with boundary conditions

$$B_1(\theta)=\theta-\theta_w=0 \qquad \text{on } a_1 \tag{6.2.2}$$

$$B_2(\theta)=k\frac{\partial\theta}{\partial n}+q_w \qquad \text{on } a_2 \tag{6.2.3}$$

$$B_3(\theta)=k\frac{\partial\theta}{\partial n}+h(\theta-\theta_f)=0 \qquad \text{on } a_3 \tag{6.2.4}$$

where θ_w is the temperature of the body surface and θ_f is the fluid temperature. We look for a solution of the problem (6.2.1) – (6.2.4) in the class of functions θ, fulfilling the boundary conditions (6.2.2) – (6.2.4). These functions in general do not fulfill the differential equation (6.2.1). Substituting one of these functions into Eq. (6.2.1) we get

$$\nabla\cdot(k\,\nabla\theta)+r-C_{\Delta}\frac{\partial\theta}{\partial t}=E\neq 0 \tag{6.2.5}$$

The best approximation for the solution will be a function θ which minimizes the residuum E. The simplest method of obtaining the solution is to make use of the fact that if E is identically equal to zero, then

$$\int_V wE\,dV=0 \tag{6.2.6}$$

where V is the volume of the body under consideration and w is an arbitrary function.

A similar procedure is used for boundary conditions. We get

$$\int_V w\left[\nabla(k\,\nabla\theta)+r-C_{\Delta}\frac{\partial\theta}{\partial t}\right]dV=0 \tag{6.2.7}$$

and

$$\sum_{i=1}^{3} \int_{a_i} w_i B_i(\theta)\, da_i = 0 \tag{6.2.8}$$

The above conditions may be replaced by the single condition

$$\int_V w_0 \left[\nabla(k\,\nabla\theta) + r - C_\Delta \frac{\partial\theta}{\partial t} \right] dV + \sum_{i=1}^{3} \int_{a_i} w_i B_i(\theta)\, da_i = 0 \tag{6.2.9}$$

Since we are dealing with a second-order partial differential operator the function θ must be continuous with continuous first-order derivatives. If we use the divergence theorem to lessen the order of differentiation in (6.2.9) then we may simply require the temperature to be of class of zero-order. Then we can express it in terms of shape functions, for which this condition is fulfilled. Using the divergence theorem, the requirements as to the function w increase; it now has to be a condition function. Making use of the divergence theorem for Eq. (6.2.9) we get

$$\begin{aligned} &\int_V \nabla w (k\,\nabla\theta) dV - \int_V wr\, dV - \int_V wC_\Delta \frac{\partial\theta}{\partial t} dV - \int_{a_i} wk \frac{\partial\theta}{\partial n} da \\ &- \int_{a_2} w_2 \left[k\frac{\partial\theta}{\partial n} + q_w \right] da - \int_{a_3} w_3 \left[k\frac{\partial\theta}{\partial n} + h(\theta - \theta_f) \right] da = 0 \end{aligned} \tag{6.2.10}$$

where $a = a_1 \cup a_2 \cup a_3$ is the total surface of the body.
Since functions w_i are arbitrary continuous functions, we can put

$$w_2 = w_3 = -w \tag{6.2.11}$$

Finally we get

$$\begin{aligned} &\int_V \nabla w\, k\, \nabla\theta\, dV - \int_V wr\, dV - \int_V wC_\Delta \frac{\partial\theta}{\partial t} dV + \int_{a_2} wq_w\, da \\ &+ \int_{a_3} wh(\theta - \theta_f) da - \int_{a_i} wk \frac{\partial\theta}{\partial n} da = 0 \end{aligned} \tag{6.2.12}$$

Assume that the temperature can be approximated by the expression

$$\theta = \sum_{i=1}^{I} H_i \theta_i \tag{6.2.13}$$

where H_i are shape functions, θ_i are nodal temperatures and I stands for the number of nodes. The expression (6.2.13) is sometimes shown in matrix notation as

$$\theta = \mathbf{H}^T \boldsymbol{\theta} \tag{6.2.14}$$

where $\mathbf{H}$ is the vector of shape functions and $\boldsymbol{\theta}$ is the vector of nodal temperatures. Functions w in Eq. (6.2.12) can be taken as

$$w = H_i \tag{6.2.15}$$

This assumption reduces the weighted residual method to the classical Galerkin method. The particular advantage of such an approach is that for heat exchange problems in non-moving bodies the matrix of the symmetric and banded.

Substitute in Eq. (6.2.12) the functions

$$w = H_i \qquad i = 1, 2, \ldots, I \tag{6.2.16}$$

By substitution of temperature θ in the form of (6.2.13) we get a system of I equations with I unknowns, θ_i

$$\mathbf{K}^\theta \boldsymbol{\theta} + \mathbf{C}^\theta \dot{\boldsymbol{\theta}} = \mathbf{F}^\theta \tag{6.2.17}$$

where

$$K_{ij}^\theta = \int_V \nabla H_i k \nabla H_j dV + \int_{a_3} H_i h H_j da - \int_{a_1} H_i k \frac{\partial H_j}{\partial n} da \tag{6.2.18}$$

$$C_{ij}^\theta = \int_V H_i C_\Delta H_j dV \tag{6.2.19}$$

$$F_i^\theta = \int_V H_i r dV - \int_{a_2} H_i q_w da + \int_{a_3} H_i h \theta_f da \tag{6.2.20}$$

As we see, the matrix $\mathbf{K}^\theta$ is symmetric. Integrals in expressions (6.2.18) and (6.2.19) are different from zero only in the sub-domains where $H_i \neq 0$, e.g. in the elements which possess the i-th node.

6.2.2 Variational Formulation

The heat exchange problem is uniquely defined by the partial differential equation with boundary conditions. It is however possible to formulate the problem using the extreme variational principle. According to Euler's theorem the solution of the partial differential equation of heat transfer

$$k_{11}\frac{\partial^2\theta}{\partial x_1^2} + k_{22}\frac{\partial^2\theta}{\partial x_2^2} + k_{33}\frac{\partial^2\theta}{\partial x_3^2} + r = C_\Delta \frac{\partial\theta}{\partial t} \tag{6.2.21}$$

may be found by minimizing the functional

$$\mathcal{H} = \frac{1}{2}\int_V \left\{ \left[k_{11}\left(\frac{\partial\theta}{\partial x_1}\right)^2 + k_{22}\left(\frac{\partial\theta}{\partial x_2}\right)^2 + k_{33}\left(\frac{\partial\theta}{\partial x_3}\right)^2 \right] - 2r\theta + C_\Delta\theta^2 + 2C_\Delta\theta_0\theta \right\} dV \tag{6.2.22}$$

with assumptions that the function θ fulfils the given boundary conditions

$$\theta = \theta_w \quad \text{on } a_1 \tag{6.2.23}$$

$$k\frac{\partial\theta}{\partial n} + q_w = 0 \quad \text{on } a_2 \tag{6.2.24}$$

$$k\frac{\partial\theta}{\partial n} + h(\theta - \theta_f) = 0 \quad \text{on } a_3 \tag{6.2.25}$$

and θ_0 is the initial temperature distribution.

The condition (6.2.23) can be easily fulfilled. However, the remaining conditions create significant difficulties. Therefore one adds to the functional (6.2.22) the surface integral, which in the process of minimization leads to (6.2.24) or (6.2.25). In general it has the form

$$\frac{1}{2}\left\{\int_{a_2} 2q_w\theta\,da + \int_{a_3} h\left(\theta^2 - 2\theta\theta_f\right)da\right\} \tag{6.2.26}$$

or

$$\overline{\mathcal{H}} = \mathcal{H} + \frac{1}{2}\left\{\int_{a_2} 2q_w\theta\,da + \int_{a_3} h\left(\theta^2 - 2\theta\theta_f\right)da\right\} \tag{6.2.27}$$

and at the same time the function θ has to fulfill the condition (6.2.23).

The functional in the form (6.2.27) can be expressed as the sum of functionals determined for each differential element

$$\mathcal{H} = \sum_n {}^n\overline{\mathcal{H}} \tag{6.2.28}$$

It reaches a minimum, if all components of the sum are minimal. Assume that the temperature θ is expressed by shape functions and nodal temperatures

$$\theta = \sum_i H_i\theta_i \tag{6.2.29}$$

Since all shape functions are continuous functions, derivatives appearing in the functional (6.2.27) are determined in order to minimize this functional with respect to parameters θ_i, we should solve the following system of equations

$$\frac{\partial\overline{\mathcal{H}}}{\partial\theta_i} = 0 \qquad \text{for } i = 1, 2, \ldots, I \tag{6.2.30}$$

Finally we get

$$\mathbf{K}^\theta\boldsymbol{\theta} + \mathbf{C}^\theta\dot{\boldsymbol{\theta}} = \mathbf{F}^\theta \tag{6.2.31}$$

where

$$K_{ij}^{\theta} = \int_V \nabla H_i k \nabla H_j dV + \int_{a_3} H_i h H_j da - \int_{a_1} H_i k \frac{\partial H_j}{\partial n} da \tag{6.2.32}$$

$$C_{ij}^{\theta} = \int_V H_i C_{\Delta} H_j dV \tag{6.2.33}$$

$$F_i^{\theta} = \int_V H_i r\, dV - \int_{a_2} H_i q_w da + \int_{a_3} H_i h \theta_f da \tag{6.2.34}$$

and at the same time the surface integrals appear only in the cases where nodes are placed on the body boundary.

6.3 Finite-Element Solution of Navier-Stokes Equations

Navier-Stokes equations (4.2.1) – (4.2.3) with boundary conditions given by (4.2.6) and (4.2.7) form a complete set for the determination of the velocity, pressure and temperature fields in a moving fluid

$$C_{\Delta}\left(\frac{\partial \theta}{\partial t} + (\mathbf{v}\nabla)\theta\right) = \nabla(k\nabla\theta) + r \tag{6.3.1}$$

$$\nabla\, \mathbf{v} = 0 \tag{6.3.2}$$

$$\rho\left[\frac{\partial \mathbf{v}}{\partial t} + \mathbf{v}(\nabla\, \mathbf{v})\right] = \nabla\, \mathbf{T} - \rho g \beta(\theta - \theta^0) \tag{6.3.3}$$

where

$$T_{ij} = -p\delta_{ij} + \mu(v_{i,j} + v_{j,i}) \tag{6.3.4}$$

In matrix notation we can write

$$\nabla\, \mathbf{T} = -\nabla p\, \mathbf{I} + \mu \nabla \mathbf{\Lambda}^T\, \mathbf{v} \tag{6.3.5}$$

To apply the method of weighted residuals to this problem, we assume that within each element, a set of nodal points is established at which the dependent variables v_i, p and θ are evaluated. It is assumed that the dependent variables can be expressed in terms of approximating functions by

$$\mathbf{v} = \tilde{\mathbf{v}} \qquad p = \tilde{p} \qquad \theta = \tilde{\theta} \tag{6.3.6}$$

where ~ indicates an approximation. To derive a discrete set of equations appropriate for a particular element, the approximating functions are expressed by

$$\tilde{\mathbf{v}} = \mathbf{N}^{\mathrm{T}}\, \mathbf{v}(t) \qquad \tilde{p} = \mathbf{M}^{\mathrm{T}}\mathbf{P}(t) \qquad \tilde{\theta} = \mathbf{H}^{\mathrm{T}}\, \boldsymbol{\theta}(t) \tag{6.3.7}$$

where **N** is a matrix, **M**, **H** are vectors of shape functions for the element and **v**, **P**, **θ** are vectors of nodal point unknowns.
By Eqs. (6.3.1) – (6.3.5) with boundary conditions (4.2.6) and (4.2.7) we get the system of finite element equations

$$\left(\int_V \mathbf{H}\, C_\Delta\, \mathbf{H}^{\mathrm{T}} \frac{\partial\theta}{\partial t} + \left(C_\Delta\, \mathbf{H}\, \mathbf{N}^{\mathrm{T}}\mathbf{v}\,\nabla\,\mathbf{H}^{\mathrm{T}}\, \boldsymbol{\theta}\right)\right) dV + \int_V \nabla\, \mathbf{H}\, k \nabla\, \mathbf{H}^{\mathrm{T}}\, \boldsymbol{\theta}\, dV$$
$$= \int_a \mathbf{H}\, \mathbf{q}\, \mathbf{n}\, da + \int_V \mathbf{H}\, r\, dV \tag{6.3.8}$$

$$\left(\int_V \mathbf{M}\,\nabla\, \mathbf{N}^{\mathrm{T}}\, dV\right)\mathbf{v} = 0 \tag{6.3.9}$$

$$\int_V \rho\left(\mathbf{N}\mathbf{N}^{\mathrm{T}}\frac{\partial\,\mathbf{v}}{\partial t}\right) dV + \left(\int_V \rho\mathbf{N}\mathbf{N}^{\mathrm{T}}\mathbf{v}\nabla\mathbf{N}^{\mathrm{T}}\, dV\right)\mathbf{v} - \left(\int_V \nabla\,\mathbf{N}^{\mathrm{T}}\mathbf{M}\, dV\right)\mathbf{P}$$
$$+\left(\int_V \mu\nabla\boldsymbol{\Lambda}^{\mathrm{T}} dV\right)\mathbf{v} + \int_V \rho g\beta\mathbf{N}\,\mathbf{H}^{\mathrm{T}}\, dV\left(\theta_r - \theta_r^0\right) + \int_a \mathbf{N}\,\boldsymbol{\tau}\,\mathbf{n}\, da = 0 \tag{6.3.10}$$

From Eqs. (6.3.8) – (6.3.10) we get

$$\mathbf{C}^{\theta}\, \dot{\boldsymbol{\theta}} + \mathbf{K}^{\theta}\, \boldsymbol{\theta} + \mathbf{L}^{\theta}\, \boldsymbol{\theta} = \mathbf{F}^{\theta} \tag{6.3.11}$$

$$\mathbf{A}^{v}\,\mathbf{v} = 0 \tag{6.3.12}$$

$$\mathbf{R}^{v}\,\dot{\mathbf{v}} + \mathbf{E}^{v}\,\mathbf{v} + \mathbf{D}^{v}\,\mathbf{v} = \mathbf{G}^{v} + \mathbf{B}^{v}\,\mathbf{P} \tag{6.3.13}$$

where

$$\mathbf{C}^{\theta} = \int_{V} C_{\Delta}\,\mathbf{H}\,\mathbf{H}^{T}\,dV \tag{6.3.14}$$

$$\mathbf{K}^{\theta} = \int_{V} \nabla\,\mathbf{H}\,k\nabla\,\mathbf{H}^{T}\,dV \tag{6.3.15}$$

$$\mathbf{L}^{\theta} = \int_{V} C_{\Delta}\,\mathbf{H}\,\mathbf{N}^{T}\mathbf{v}\nabla\,\mathbf{H}\,dV \tag{6.3.16}$$

$$\mathbf{F}^{\theta} = \int_{V} \mathbf{H}\,r\,dV - \int_{a} \mathbf{H}\,\mathbf{q}\,\mathbf{n}\,da \tag{6.3.17}$$

$$\mathbf{A}^{v} = \int_{V} \mathbf{M}\nabla\,\mathbf{N}^{T}\,dV \tag{6.3.18}$$

$$\mathbf{R}^{v} = \int_{V} \rho\,\mathbf{N}\,\mathbf{N}^{T}\,dV \tag{6.3.19}$$

$$\mathbf{E}^{v} = \int_{V} \rho\,\mathbf{N}\,\mathbf{N}^{T}\mathbf{v}\nabla\,\mathbf{N}^{T}\,dV \tag{6.3.20}$$

$$\mathbf{D}^{v} = \mu\int_{V} \nabla\boldsymbol{\Lambda}^{T}\,\mathbf{N}\,dV \tag{6.3.21}$$

$$\mathbf{G}^{V} = -\int_{V} \rho g \beta \mathbf{N}\,\mathbf{H}^{T}\,dV\left(\theta - \theta^{0}\right) + \int_{a} \mathbf{N}\,\tau\,\mathbf{n}\,da \tag{6.3.22}$$

$$\mathbf{B}^{V} = \int_{V} \nabla \mathbf{N}^{T} \mathbf{M}\,dV \tag{6.3.23}$$

6.4 Time Discretization

Assume the incremental approach to the equations describing the welding process. Thus this approach is applied to equations of thermo-elasto-plasticity described in Section 5.9, equations of heat flow given in Section 6.2 and Navier-Stokes equations presented in Section 6.3. The time discretization gives

$$t_n = n \cdot \Delta t \tag{6.4.1}$$

where n is the time step.

Starting from the solution S_{n-1}

$$S_{n-1} = (\mathbf{T}_{n-1}, \mathbf{u}_{n-1}, \boldsymbol{\theta}_{n-1}, \mathbf{v}_{n-1}) \tag{6.4.2}$$

known at time t_{n-1} the solution S_n at time t_n needs to be derived. Formally

$$S_n = S_{n-1} + \Delta_n S \qquad \text{where} \qquad \Delta_n S = (\Delta_n \mathbf{T}, \Delta_n \mathbf{u}, \Delta\boldsymbol{\theta}_{n-1}, \Delta\mathbf{v}_{n-1}) \tag{6.4.3}$$

The problem discretized with respect to time satisfies the discretized momentum equation, heat flow equation, the compatibility equation and the discretized thermal and mechanical boundary conditions.

The discretized equations and their solutions are discussed in the subsequent sections.

6.5 Time Integration Schemes for Nonlinear Heat Conduction

In this section we discuss various algorithms that have been proposed for numerical integration after finite-element discretization. The algorithms reviewed comprise one-step and two-step schemes. The study is an attempt to define criteria for an optimum choice among such algorithms, where emphasis is given to the accuracy achievable.

Consider the non-linear transient heat transfer matrix equations (6.2.17) – (6.2.20)

$$\mathbf{K}^{\theta}(\boldsymbol{\theta})\boldsymbol{\theta}(t)+\mathbf{C}^{\theta}(\boldsymbol{\theta})\dot{\boldsymbol{\theta}}(t)=\mathbf{F}^{\theta}(\boldsymbol{\theta},t)$$

$$\boldsymbol{\theta}(0)=\boldsymbol{\theta}_0 \qquad (6.5.1)$$

where $\boldsymbol{\theta}$ (t) is the vector of the nodal temperatures (with a superscript dot denoting its time derivative), $\mathbf{K}^{\theta}$ is the symmetrical and positive semi-definite conductivity matrix, $\mathbf{C}^{\theta}$ is the symmetrical and positive definite capacity matrix and $\mathbf{F}^{\theta}$ is a vector of thermal loads corresponding to $\boldsymbol{\theta}$; $\boldsymbol{\theta}_0$ is a vector of given initial temperatures. In this section we shall consider and compare some of the methods currently used and proposed for the second stage of this solution procedure, i.e. the time integration of system (6.5.1). Thus we leave completely apart the problem of iterative solution of the resulting set of non-linear algebraic equations.

In order to undertake such a review and to facilitate the comparison with computational techniques applied to typical first-order differential equations, we rewrite system (6.5.1) in the form

$$\dot{\boldsymbol{\theta}}(t)=-\mathbf{A}^{\theta}(\boldsymbol{\theta})\boldsymbol{\theta}(t)+\mathbf{D}^{\theta}(\boldsymbol{\theta},t)=\mathbf{G}^{\theta}(\boldsymbol{\theta},t)$$

$$\boldsymbol{\theta}(0)=\boldsymbol{\theta}_0 \qquad (6.5.2)$$

where $\mathbf{A}^{\theta}=\mathbf{C}^{\theta^{-1}}\mathbf{K}^{\theta}$ has real positive eigenvalues and $\mathbf{D}^{\theta}=\mathbf{C}^{\theta^{-1}}\mathbf{F}^{\theta}$ is a new forcing vector. Such a system is a stiff one, i.e. composed of exponentially decaying components with widely spread time constants which, in addition, change continuously in the non-linear case.

In welding applications one deals with situations where the slowly decaying components dominate the response; hence attention will be focused on time integration techniques that treat adequately the long-term components of the response while retaining numerical stability with respect to the fast-varying excitations. The simplest kind of time integration schemes are one-step schemes in which a two-level difference approximation is chosen for $\dot{\boldsymbol{\theta}}$ and a linear variation of $\mathbf{G}^{\theta}$ is assumed over the interval $[t_n, t_{n+1}]$, thus yielding

$$\boldsymbol{\theta}_{n+1}-\boldsymbol{\theta}_n=\Delta t\left[(1-\vartheta)\mathbf{G}_n^{\theta}+\vartheta\mathbf{G}_{n+1}^{\theta}\right] \qquad 0\le\vartheta\le 1 \qquad (6.5.3)$$

where Δt is the time step and $\mathbf{G}_n^{\theta}$ stands for $\mathbf{G}^{\theta}(\boldsymbol{\theta}_n, n\Delta t)$.

These methods will be referred to as generalized trapezoidal schemes. A slightly different form of one-step scheme is obtained from the same two-level difference approximation for $\dot{\boldsymbol{\theta}}$ and from an assumed linear behaviour of $\boldsymbol{\theta}$ over $[t_n, t_{n+1}]$

$$\boldsymbol{\theta}_{n+1} - \boldsymbol{\theta}_n = \Delta t\, \mathbf{G}^{\theta}_{\vartheta} = \Delta t\, \mathbf{G}^{\theta}(\boldsymbol{\theta}_{\vartheta}, t_{\vartheta}) \tag{6.5.4}$$

in which

$$\begin{aligned} &\boldsymbol{\theta}_{\vartheta} = (1-\vartheta)\boldsymbol{\theta}_n + \vartheta\boldsymbol{\theta}_{n+1} \\ & \qquad\qquad\qquad\qquad\qquad\qquad 0 \le \vartheta \le 1 \\ &t_{\vartheta} = (1-\vartheta)t_n + \vartheta t_{n+1} = t_n + \vartheta\Delta t \end{aligned} \tag{6.5.5}$$

They will be referred to as generalized mid-point schemes. the two families reduce to one if **G** is a linear function of the unknown (e.g. linear heat conduction): the so-called 'ϑ-method'. Particular cases are well known

$$\vartheta = \begin{cases} 0 & \text{Euler explicit (forward) scheme,} \\ \frac{1}{2} & \text{either trapezoidal or Crank - Nicolson (mid - point rule) scheme,} \\ \frac{2}{3} & \text{Galerkin scheme,} \\ 1 & \text{fully implicit or Euler backward scheme.} \end{cases}$$

All the schemes are consistent ones of the first order, except for $\vartheta = \frac{1}{2}$, in which case the two families are second-order accurate in the time step size.

Applied to system (6.5.2), the schemes (6.5.3) and (6.5.4) yield the following set of nonlinear algebraic equations, with **I** denoting the identity matrix

$$\left(\mathbf{I} + \vartheta\Delta t\, \mathbf{A}^{\theta}_{n+1}\right)\boldsymbol{\theta}_{n+1} = \left[\mathbf{I} - (1-\vartheta)\Delta t\, \mathbf{A}^{\theta}_{n}\right]\boldsymbol{\theta}_n + (1-\vartheta)\Delta t\, \mathbf{D}^{\theta}_{n} + \vartheta\Delta t\, \mathbf{D}^{\theta}_{n+1} \tag{6.5.6}$$

for trapezoidal schemes

$$\left(\mathbf{I} + \vartheta\Delta t\, \mathbf{A}^{\theta}_{\vartheta}\right)\boldsymbol{\theta}_{n+1} = \left[\mathbf{I} - (1-\vartheta)\Delta t\, \mathbf{A}^{\theta}_{\vartheta}\right]\boldsymbol{\theta}_n + \Delta t\, \mathbf{D}^{\theta}_{\vartheta} \tag{6.5.7}$$

for mid-point schemes.

Thus every step involves the construction and solution of a new set of equations; therefore iterative solution techniques (except for $\vartheta = 0$) are required in which tangent or secant approximations may be used.

A straight linearization has been proposed over a time step by assuming

$$\mathbf{A}^{\theta}_{n+1} \cong \mathbf{A}^{\theta}_{n} \qquad \mathbf{D}^{\theta}_{n+1} \cong \mathbf{D}^{\theta}(\boldsymbol{\theta}_n, t_{n+1}) \tag{6.5.8}$$

for trapezoidal schemes
or

$$\mathbf{A}^{\theta}_{\vartheta} \cong \mathbf{A}^{\theta}_{n} \qquad \mathbf{D}^{\theta}_{\vartheta} \cong \mathbf{D}^{\theta}(\boldsymbol{\theta}_n, t_{\vartheta}) \tag{6.5.9}$$

for mid-point schemes. This corresponds obviously to a secant or initial load method of solution of the non-linear systems (6.5.6) or (6.5.7).

An extrapolation technique for the mid-point schemes family has been proposed by taking

$$\boldsymbol{\theta}_{\vartheta} = (1+\vartheta)\boldsymbol{\theta}_n - \vartheta\boldsymbol{\theta}_{n-1} \tag{6.5.10}$$

A predictor-corrector technique has been proposed for the mid-point schemes family in which a predicted value $\boldsymbol{\theta}^*_{n+1}$ stems from (6.5.7), by assuming (6.5.9); a corrected value is then obtained by again solving (6.5.7) with

$$\boldsymbol{\theta}^*_{\vartheta} = (1-\vartheta)\boldsymbol{\theta}_n + \vartheta\boldsymbol{\theta}^*_{n+1}$$

The procedure thus requires the solution of two linear systems of equations per time step.

In two-step methods we consider the expression

$$\begin{aligned}&\left(\vartheta+\tfrac{1}{2}\right)\boldsymbol{\theta}_{n+2} - 2\vartheta\boldsymbol{\theta}_{n+1} + \left(\vartheta+\tfrac{1}{2}\right)\boldsymbol{\theta}_n + \Delta t\,\mathbf{A}^{\theta}_{*}\left[\beta_2\boldsymbol{\theta}_{n+2}\right.\\ &\left.+ (1+\vartheta-2\beta_2)\boldsymbol{\theta}_{n+1} + (\beta_2-\vartheta)\boldsymbol{\theta}_n\right] = \Delta t\,\mathbf{D}^{\theta}_{*}\end{aligned} \tag{6.5.11}$$

Clearly every step implies the construction and solution of a new system of linear equations with the positive definite system matrix $\left[\left(\tfrac{1}{2}+\vartheta\right)\mathbf{I} + \beta_2\Delta t\,\mathbf{A}^{\theta}_{*}\right]$. A different starting procedure is needed since the first application of (6.5.11) requires knowledge of $\boldsymbol{\theta}_1$. One of the previous one-step schemes can be used for this purpose, but the problems associated with that changeover (e.g. step length change) will not be considered here.

Table 6.1 shows how most of the two-step schemes that have been presented in the literature are derived from (6.5.11) by particular choice of ϑ and β_2. It should be noted that some of these schemes are equivalent to particular one-step schemes: this is the case for the Dupont I scheme (Dupont et al. 1974) when $\alpha = \tfrac{1}{2}$ and the Dupont II scheme (6.5.15) when $\alpha = 0$, that both restore the Crank – Nicolson algorithm (scheme (6.5.7) with

$\vartheta = \frac{1}{2}$). In addition a linearization of the unknown with [t_n, t_{n+2}], i.e. the assumption that

$$\boldsymbol{\theta}_{n+1} = \left(\boldsymbol{\theta}_{n+2} + \boldsymbol{\theta}_n\right)/2 \tag{6.5.12}$$

also restores particular one-step schemes. These are known as the Crank–Nicolson scheme for (6.5.13) and (6.5.14) for whatever the value of α, the mid-point scheme (6.5.7) with $\vartheta = \frac{3}{4}$ for (6.5.15) for whatever the value of α, and the fully implicit mid-point scheme ($\vartheta = 1$) for (6.5.16).

The two-step methods are widely used in solutions of problems involving phase change.

Table 6-1. Linearized two-step schemes

ϑ	β_2	$\boldsymbol{\theta}_*$	t_*	Algorithm name	Integration scheme	
0	$\frac{1}{3}$	$\boldsymbol{\theta}_{n+1}$	t_{n+1}	Lee	$\left(\frac{1}{2}\mathbf{I} + \frac{1}{3}\Delta t\,\mathbf{A}^{\theta}_{n+1}\right)\boldsymbol{\theta}_{n+2} = -\frac{1}{3}\Delta t\,\mathbf{A}^{\theta}_{n+1}\,\boldsymbol{\theta}_{n+1} + \left(\frac{1}{2}\mathbf{I} - \frac{1}{3}\Delta t\,\mathbf{A}_{n+1}\right)\boldsymbol{\theta}_n + \Delta t\,\mathbf{D}^{\theta}_{n+1}$	(6.5.13)
0	α $\left(\alpha > \frac{1}{4}\right)$	$\boldsymbol{\theta}_{n+1}$	t_{n+1}	Dupont I	$\left(\frac{1}{2}\mathbf{I} + \alpha\Delta t\,\mathbf{A}^{\theta}_{n+1}\right)\boldsymbol{\theta}_{n+2} = -(1-2\alpha)\Delta t\,\mathbf{A}^{\theta}_{n+1}\,\boldsymbol{\theta}_{n+1} + \left(\frac{1}{2}\mathbf{I} - \alpha\Delta t\,\mathbf{A}^{\theta}_{n+1}\right)\boldsymbol{\theta}_n + \Delta t\,\mathbf{D}^{\theta}_{n}$	(6.5.14)
$\frac{1}{2}$	$\frac{1}{2} + \alpha$ $(\alpha > 0)$	$\frac{3}{2}\boldsymbol{\theta}_{n+1} - \frac{1}{2}\boldsymbol{\theta}_n$	$t_{n+\frac{3}{2}}$	Dupont II	$\left(\mathbf{I} + \left(\frac{1}{2} + \alpha\right)\Delta t\,\mathbf{A}^{\theta}_*\right)\boldsymbol{\theta}_{n+2} = \left[\mathbf{I} - \left(\frac{1}{2} - 2\alpha\right)\Delta t\,\mathbf{A}^{\theta}_*\right]\boldsymbol{\theta}_{n+1} - \alpha\Delta t\,\mathbf{A}^{\theta}_*\,\boldsymbol{\theta}_n + \Delta t\,\mathbf{D}^{\theta}_*$	(6.5.15)
1	1	$2\boldsymbol{\theta}_{n+1} - \boldsymbol{\theta}_n$	t_{n+2}	Linearized fully implicit	$\left(\frac{3}{2}\mathbf{I} + \Delta t\,\mathbf{A}^{\theta}_*\right)\boldsymbol{\theta}_{n+2} = 2\boldsymbol{\theta}_{n+1}$ $\frac{1}{2}\boldsymbol{\theta}_n + \Delta t\,\mathbf{D}^{\theta}_*$	(6.5.16)

6.6 Solution Procedure for Navier-Stokes Equation

Thermal field and fluid flow field can be found using the factorial step method. With the solution known everywhere at time t_n we determine the solution of the fluid flow equation at the next time level t_{n+1}. In this method an intermediate velocity field $\mathbf{v}^*$ is calculated to satisfy only a discretized version of Eq. (6.3.13) without the pressure term. We have

$$\Delta\,\mathbf{v} = \Delta t\,\mathbf{R}^{v^{-1}}\left(\mathbf{G}^{v} - \mathbf{E}^{v}\,\mathbf{v}_n - \mathbf{D}^{v}\,\mathbf{v}_n\right) \tag{6.6.1}$$

$$\mathbf{v}^* = \mathbf{v}_n + \Delta\,\mathbf{v} \qquad (6.6.2)$$

The required velocity and pressure fields $\mathbf{v}_{n+1}$ and $\mathbf{P}_{n+1}$ are then obtained by adding to $\mathbf{v}^*$ the dynamic effect of pressure determined to ensure that the incompressibility condition remains satisfied

$$\Delta t\, \mathbf{E}\, \mathbf{R}^{v^{-1}}\, \mathbf{E}^{\,v} \mathbf{P}_{n+1} = -\mathbf{E}^{v}\, \mathbf{v}^* \qquad (6.6.3)$$

With the velocity and pressure at time t_{n+1} so determined the solution is completed by evaluating the temperature at this time from Eq. (6.3.11).

$$\Delta\boldsymbol{\theta} = \Delta t\, \mathbf{C}^{\theta^{-1}} \left(\mathbf{F}^{\theta} - \mathbf{K}^{\theta}\, \boldsymbol{\theta} - \mathbf{L}^{\theta}\, \boldsymbol{\theta}\right) \qquad (6.6.4)$$

$$\boldsymbol{\theta}_{n+1} = \boldsymbol{\theta}_n + \Delta\boldsymbol{\theta} \qquad (6.6.5)$$

6.7 Modeling of the Phase Change Process

Two major problems make the welding stress analysis of materials involving phase change difficult to handle. The first is the method of accounting for latent heat absorption or release during the phase change. The second one is the kinematics of the phase boundary. The abrupt change of properties during phase change (Fig. 6-1) for most materials undoubtedly causes numerical instability in the analysis.

Attempts have been made to solve this type of problem by the classical method using the heat conduction equation with the latent heat and moving solid-liquid interface included in the boundary conditions. But only a limited number of problems involving simple geometries can be solved in this way. A different concept is the so-called enthalpy method.

In constructing the solution of problems involving phase change a possible approach is to track accurately the position of the phase boundary and then to solve Eq. (6.2.1) for the solid region and Eqs. (6.3.1) – (6.3.5) for the fluid region. Consider the case when the phase change process is modeled by a variant of the enthalpy method. In this method the phase change is assumed to occur over a temperature range and the associated latent heat effects is handled by increasing suitably the specific heat in this range. Thus if the phase change is assumed to occur over the temperature interval $[\theta_s, \theta_l]$, where θ_l is the liquidus and θ_s the solidus temperature, then

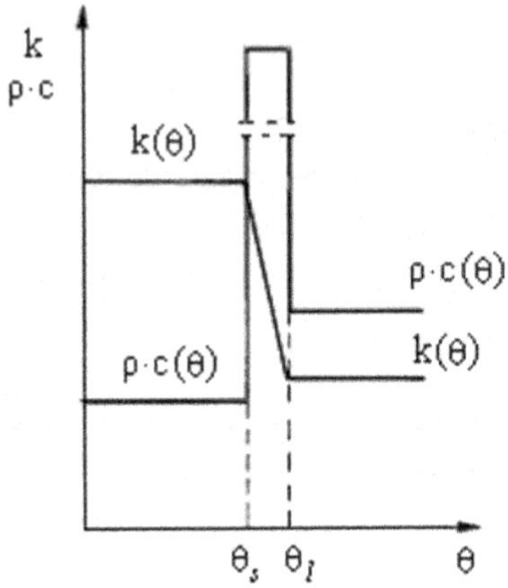

Figure 6-1. Thermomechanical properties in the range of phase change

the specific heat c_L used in the calculation is defined by

$$c_L = c + \frac{[H(\theta-\theta_s) - H(\theta-\theta_l)]L}{\Delta\theta} \tag{6.7.1}$$

where H denotes the Heaviside function

$$H(t-a) = \begin{cases} 1 & (t > a) \\ 0 & (t \le a) \end{cases} \tag{6.7.2}$$

L is the latent heat, and $\Delta\theta = \theta_l - \theta_s$ is the phase change interval.

When this method is applied to the analysis of pure materials, in which the phase change occurs at a specified temperature (i.e. $\theta_s = \theta_l$), a phase change time interval Δt ($\neq 0$) must be assumed. It has been demonstrated that reasonable results can be obtained for problems involving conduction, provided that the right choice is made for the values of $\Delta\theta$. However, for materials in which the phase change does occur over a reasonable temperature range this problem does not arise and the actual physical values of θ_s and θ_l can be used successfully. The enthalpy method can be used in conjunction with a fixed finite-element mesh. It should be noted that such an approach was well suited to the computation of fluid motion except for the problem of the application of the non-slip (or zero fluid-velocity) boundary condition at the phase boundary. Exact application of this boundary condition at the phase change surface will not be possible as the fluid-solid interface will, in general, not coincide with the nodes of the finite-element mesh.

We overcame this problem by using a smearing approach in which the viscosity was greatly increased in the phase change region. The problem can be also overcome by adopting the simpler, but perhaps less accurate approach of reducing the nodal velocity to zero whenever the nodal temperature lies below the liquidus.

6.8 The Theorem of Virtual Work in Finite Increments

Consider a field of finite stress increments $\Delta_n \mathbf{T}$ that satisfies momentum equation

$$\operatorname{div} \Delta_n \mathbf{T} = 0 \tag{6.8.1}$$

with relation $\Delta_n \mathbf{T} \cdot \mathbf{n} = \Delta_n \mathbf{t}$ on the boundary a of the considered domain V.

Let $\Delta_n \mathbf{u}^*$ and $\Delta_n \boldsymbol{\varepsilon}^*$ be respectively, a field of virtual finite increments of material displacement and a field of virtual strain finite increments which are linked through compatibility relation

$$2\Delta_n \boldsymbol{\varepsilon} = \operatorname{grad} \Delta_n \mathbf{u} + (\operatorname{grad} \Delta_n \mathbf{u})^T \tag{6.8.2}$$

A theorem of virtual work in finite stress increments $\Delta_n \mathbf{T}$ and in virtual finite displacement increments $\Delta_n \mathbf{u}^*$ can be derived from the time-discretized equations in a similar way to the derivation of the theorem of virtual work (2.5.9) from the non-discretized equations of the problem

$$\int_V \Delta_n \mathbf{T} \Delta_n \boldsymbol{\varepsilon}^* \, dV - \int_a \Delta_n \mathbf{t} \, \Delta_n \mathbf{u}^* \, da = 0 \tag{6.8.3}$$

In (6.8.3) $\Delta_n \mathbf{u}$ has the dimension of a displacement, and the integrals in Eq. (6.8.3) represent work quantities. Consider two fields of kinematically admissible finite displacement increments $\Delta_n \mathbf{u}$ and $\Delta_n \mathbf{u}^o$. The difference $\Delta_n \mathbf{u} - \Delta_n \mathbf{u}^o$ is kinematically admissible with zero finite displacement increments imposed on the boundary a. By (6.8.3) applied to these fields we get

$$\int_V \Delta_n \mathbf{T} \left(\Delta_n \boldsymbol{\varepsilon} - \Delta_n \boldsymbol{\varepsilon}^o\right) dV - \int_a \Delta_n \mathbf{t} \left(\Delta_n \mathbf{u} - \Delta_n \mathbf{u}^o\right) da = 0 \tag{6.8.4}$$

for every statically admissible $\Delta_n \boldsymbol{\sigma}$ and $\Delta_n \mathbf{u}$ and kinematically admissible $\Delta_n \mathbf{u}^o$. Consider the statically admissible finite increment fields of stress $\Delta_n \mathbf{T}$ and $\Delta_n \mathbf{T}^o$. Their differences $\Delta_n \mathbf{T}$ and $\Delta_n \mathbf{T}^o$ are also statically admissible with zero data imposed on a. By (6.8.3) applied to fields of finite increments

$$\Delta_n \mathbf{u}^* = \Delta_n \mathbf{u} - \Delta_n \mathbf{u}^o \tag{6.8.5}$$

and to fields of auto-equilibrated stress ($\Delta_n \mathbf{T} - \Delta_n \mathbf{T}^o$)

$$\int_V \left(\Delta_n \mathbf{T} - \Delta_n \mathbf{T}^o\right)\left(\Delta_n \boldsymbol{\varepsilon} - \Delta_n \boldsymbol{\varepsilon}^o\right) dV = 0 \tag{6.8.6}$$

for every statically admissible $\Delta_n \mathbf{T}$, $\Delta_n \mathbf{T}^o$ and kinematically admissible $\Delta_n \mathbf{u}$, $\Delta_n \mathbf{u}^o$. By expression (6.8.6) the sum of the virtual works of auto-equilibrated finite increments of all the forces, which are developed in fields of virtual displacements kinematically admissible with zero data, is zero.

6.9 The Thermo-Elasto-Plastic Finite Element Model

Consider the virtual work equation for a finite element assemblage for a thermo-elastic-plastic material model at time $t + \Delta t$ (step n+1)

$$\int_V \mathbf{B}^T \mathbf{T}_{n+1}\, dV = \mathbf{R}_{n+1} \tag{6.9.1}$$

$$\mathbf{T}_{n+1} = \mathbf{C}_{n+1}\left(\boldsymbol{\varepsilon}_{n+1} - \boldsymbol{\varepsilon}^p_{n+1} - \boldsymbol{\varepsilon}^\theta_{n+1}\right) \tag{6.9.2}$$

$$\boldsymbol{\varepsilon}^p_n = \Lambda_n \mathbf{D} \mathbf{T}_n \tag{6.9.3}$$

$$\boldsymbol{\varepsilon}^\theta_{n+1} = \alpha\left(\theta_{n+1} - \theta_R\right)\boldsymbol{\delta} \tag{6.9.4}$$

where

$$\boldsymbol{\varepsilon}_{n+1} = \mathbf{B}\mathbf{U}_{n+1} \tag{6.9.5}$$

and $\mathbf{B}$ is the total strain-displacement transformation matrix, $\mathbf{U}_{n+1}$ is the nodal point displacement vector, $\mathbf{R}_{n+1}$ is nodal point external load vector, $\mathbf{D}$ is the deviatoric stress operator matrix and $\boldsymbol{\delta}^T$ is [1,1,1,0,0,0]. Substituting Eqs. (6.9.2) and (6.9.5) into Eq. (6.9.1) we get

$$\mathbf{K}\mathbf{U}_{n+1} = \mathbf{R}_{n+1} + \int_V \mathbf{B}^T \mathbf{C}_{n+1}\left(\boldsymbol{\varepsilon}^p + \boldsymbol{\varepsilon}^\theta\right) dV \tag{6.9.6}$$

where

$$\mathbf{K} = \int_V \mathbf{B}^T \mathbf{C}_{n+1} \mathbf{B} \, dV \tag{6.9.7}$$

is the elastic stiffness matrix.

6.10 Solution Procedure for Thermo-Elasto-Plastic Problems

The solution of thermo-elasto-plastic problems in welding using the finite-element method is a complicated process. Certain numerical difficulties have been apparent for some classes of problem, particularly when gross section yielding has been involved. The method of dealing with the constitutive equations governing the material behaviour is very important. The original formulations concentrated on forward Euler schemes, but more recently the advantages of backward Euler schemes and the radial return algorithms have been realized. For incremental finite-element analysis, the previous equations of state have to be processed within each time step t_n during load incrementation and iteration. Given the values of stress, strain, displacement and temperature at t_n, a vector of nodal loads is used to find displacements and total strains at t_{n+1}. From these, using the constitutive law which expresses the plasticity rate equations, the stress and plastic strains can be evaluated to furnish a complete solution at t_{n+1}. Load residuals may be calculated to assess the current state of convergence and then to decide whether further iterations are required before completing the current load increment.

From the algorithmic point of view, the way in which $\dot{\boldsymbol{\varepsilon}}^p$ is calculated is extremely important. Using the ϑ-method we have

$$\Delta\boldsymbol{\varepsilon}^p = \Delta t \dot{\boldsymbol{\varepsilon}}^p_{n+\vartheta} = \Delta t \lambda_{n+\vartheta} \mathbf{D} \, \mathbf{T}_{n+\vartheta} \tag{6.10.1}$$

where

$$\mathbf{T}_{n+\vartheta} = (1-\vartheta)\mathbf{T}_n + \vartheta \mathbf{T}_{n+1} \tag{6.10.2}$$

$$\lambda_{n+\vartheta} = \lambda_{n+\vartheta}\left(\mathbf{T}_{n+\vartheta}, \dot{\bar{\varepsilon}}_{n+\vartheta}, \dot{\theta}_{n+\vartheta}, \ldots\right) \tag{6.10.3}$$

and

$$\mathbf{X}_{n+\vartheta} = \mathbf{X}(\cdot, t + \vartheta\Delta t) \qquad \mathbf{X}_n = \mathbf{X}(\cdot, t) \qquad \mathbf{X}_{n+1} = \mathbf{X}(\cdot, t + \Delta t) \qquad (6.10.4)$$

for arbitrary **X**.

The values of ϑ decide which type of procedure is relevant. Three main categories exist, differing by the direction of the flow vector at current time values thereby affecting accuracy and stability. The first method is called tangent stiffness, radial corrector where $\vartheta = 0$. The second one it is mean normal where $\vartheta = \frac{1}{2}$ and the last one elastic predictor, radial return where $\vartheta = 1$.

The tangent stiffness-radial corrector ($\vartheta = 0$) method is the original method and has been used extensively. The mean normal ($\vartheta = \frac{1}{2}$) method uses a mean plastic flow direction, and conceptually lies between methods 1 and 3. The elastic predictor-radial return ($\vartheta = 1$) algorithm assumes the stress can be corrected to lie on the yield surface by applying a fixed scaling factor to all components of deviatoric stress. The radial return method, has received much attention in recent years because of the high accuracy and quadratic convergence rate when used in conjunction with a consistent tangent stiffness matrix. The value of ϑ indicates the character of the method of numerical integration. Thus $\vartheta > 0$ implies an implicit method. For Huber-Mises criterion the method is unconditionally stable if $\vartheta < \frac{1}{2}$ but unconditionally stable otherwise.

An important factor in accumulating the state variables over iterations within a load increment is the point of reference of the data. Given an increment $\Delta t = t_{n+1} - t_n$, the plasticity algorithms produce in each time-step updates for displacements, total strains, stresses and plastic strains in that order. If each is a straight update of the values at t_n, then path-dependent updating takes place. If, however, all such updates are referred back to the start of the load increment, say t_N, with each value such as dT reflecting the rate value from that time, path-independent updating occurs.

Path dependence was invariably used with the older, radial correction type algorithms but because of the somewhat oscillatory and self-compensating behaviour of the increments in the main field values particularly over the early iterations of each load step, unfortunate effects could occur. Because plasticity is a discontinuous process, Gauss points which were nearly plastic had stresses oscillating into and out of a yielded state. The unnatural unloading associated with this has the effect of producing local point-wise divergence that could grow over several iterations to render a completely divergent solution.

Part III

HEAT FLOW IN WELDING

Chapter 7

ANALYTICAL SOLUTIONS OF THERMAL PROBLEMS IN WELDING

7.1 Introduction

The solution of heat flow equations for welding conditions is a complicated problem. In a realistic model the thermal conductivity and specific heat should be considered as a functions of temperature. In most types of welding melting occurs and convective heat transfer in addition to conductive heat transfer takes place. In general, the source of heat is not concentrated at a point or line but is spread out over the workpiece with an unknown distribution of heat input. In addition, most realistic welding problems have heat losses at the boundaries caused by convection, radiation and contact with other bodies, so the precise boundary conditions are often unknown. In order to find analytical solutions to the equations, it is therefore necessary to make many simplifying assumptions. The purpose of this chapter is to present analytical solutions to the heat transfer equations under conditions of interest in welding. The restrictive assumptions will however limit the practical utility of the results. Nevertheless the results are useful in that they emphasize the variables involved and approximate the way in which they are related. In addition, such solutions provide the background for understanding more complicated solutions obtained numerically and provide guidance in making judgments.

7.2 Heat Flow in Spot Welding

Consider the welding process in the homogeneous isotropic material with constant thermal conductivity k. The heat conduction equation (3.9.4) reduces to

$$C_\Delta \frac{\partial \theta}{\partial t} = k\left(\frac{\partial^2 \theta}{\partial x_1^2} + \frac{\partial^2 \theta}{\partial x_2^2} + \frac{\partial^2 \theta}{\partial x_3^2}\right) + r \qquad (7.2.1)$$

For many practical problems it is convenient to use a system of coordinates other than Cartesian. The heat conduction in cylindrical coordinates r, ϑ, z has the form

$$\alpha\left(\frac{\partial^2 \theta}{\partial r^2} + \frac{1}{r}\frac{\partial \theta}{\partial r} + \frac{1}{r^2}\frac{\partial^2 \theta}{\partial \vartheta^2} + \frac{\partial^2 \theta}{\partial z^2}\right) + \frac{r}{C_\Delta} = \frac{\partial \theta}{\partial t} \qquad (7.2.2)$$

and in spherical coordinates r, ϑ, ϕ

$$\frac{1}{r}\frac{\partial^2 (r\theta)}{\partial r^2} - \frac{1}{r^2 \sin\vartheta}\frac{\partial}{\partial \vartheta}\left(\sin\vartheta \frac{\partial \theta}{\partial \vartheta}\right) + \frac{1}{r^2 \sin\vartheta}\frac{\partial^2 \theta}{\partial \phi^2} + \frac{r}{k} = \frac{1}{\alpha}\frac{\partial \theta}{\partial t} \qquad (7.2.3)$$

where the coefficient of thermal diffusion $k/C_\Delta = \alpha$ is introduced.

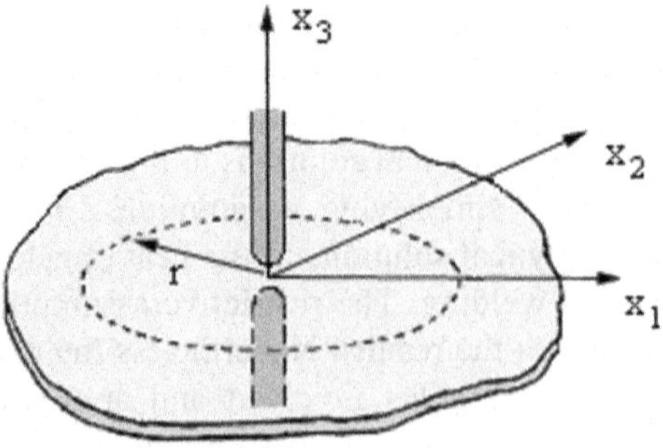

Figure 7-1. Scheme of spot welding

Thermal problems in spot welding will be solved by considering stationary line source in the middle of a large thin plate. The idealization of a spot weld is shown in Fig. 7-1. Consider the current as passing through a cylinder and consider it as a line between the two electrodes i.e. a line source along the x_3 axis of length g. Assume the current to pass through the material in zero time an instantaneous stationary line source along the x_3 axis.

The plates being welded are assumed to be of the same material and at some reference temperature equal to zero prior to the passage of the current.

Consider two dimensional case of heat flow equation and cylindrical coordinate system (r, ϑ, t).
The initial condition is

$$\theta(r, \vartheta, t_0) = \theta(r, \vartheta, 0) = 0 \tag{7.2.4}$$

The boundary conditions are

$$\left. \frac{\partial}{\partial r} \theta(r, \vartheta, t) \right|_{r=\infty} = 0 \tag{7.2.5}$$

We have

$$\int_0^\infty -2\pi k r \frac{\partial \theta}{\partial r} dt = \frac{Q}{g} \tag{7.2.6}$$

where Q is the heat input and g is the thickness of the plate.
The solution of Eq. (7.2.1) under these conditions is

$$\theta_s = \frac{Q}{g} \frac{1}{4\pi k t} \exp\left(\frac{-r^2}{4\alpha t} \right) \tag{7.2.7}$$

where the introduced subscript s refers to spot welding.
It is convenient to write Eq. (7.2.7) in dimensionless form as

$$\bar{\theta}_s = \frac{1}{\bar{t}} \exp\left(\frac{-\bar{r}_s^2}{\bar{t}} \right) \tag{7.2.8}$$

where

$$\begin{aligned} \bar{t} &= 4\alpha t / g^2 \\ \bar{r}_s &= r/g \\ \bar{\theta}_s &= \theta \pi g^3 C_\Delta / Q \end{aligned} \tag{7.2.9}$$

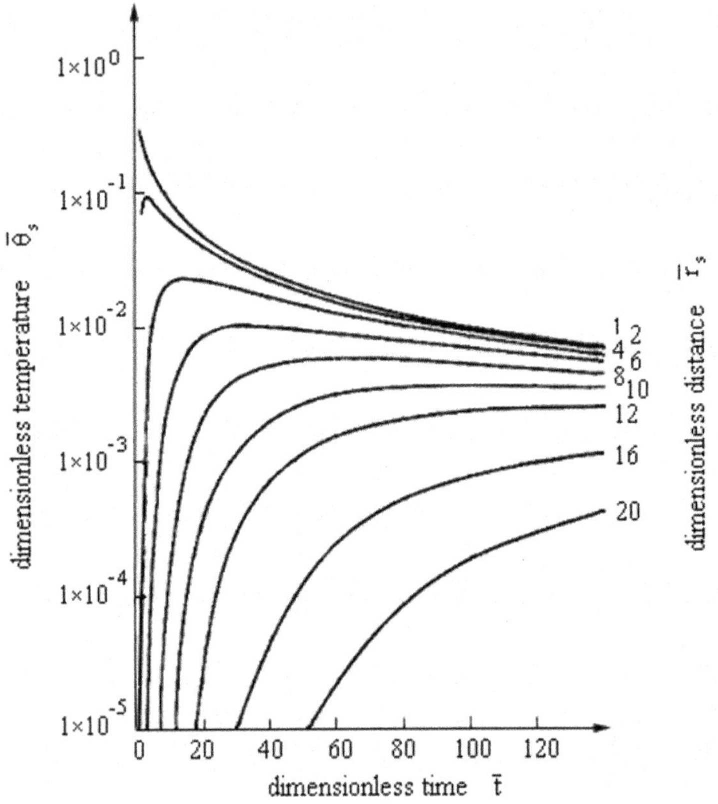

Figure 7-2. Dimensionless temperature as a function of dimensionless time for chosen constant dimensionless distances

The rate of change of temperature with time is obtained by differentiating Eq. (7.2.7) with respect to time

$$\dot{\bar{\theta}}_s = \frac{\partial \bar{\theta}_s}{\partial \bar{t}} = \frac{\exp\left(\dfrac{-\bar{r}_s^2}{\bar{t}}\right)}{\bar{t}^2}\left(\frac{\bar{r}_s^2}{\bar{t}^2} - 1\right) \tag{7.2.10}$$

The rate of heating and cooling of the metal is interesting since this affects metallurgical transformations. Fig. 7-2 illustrates dimensionless temperature as a function of dimensionless time for chosen constant dimensionless distances.

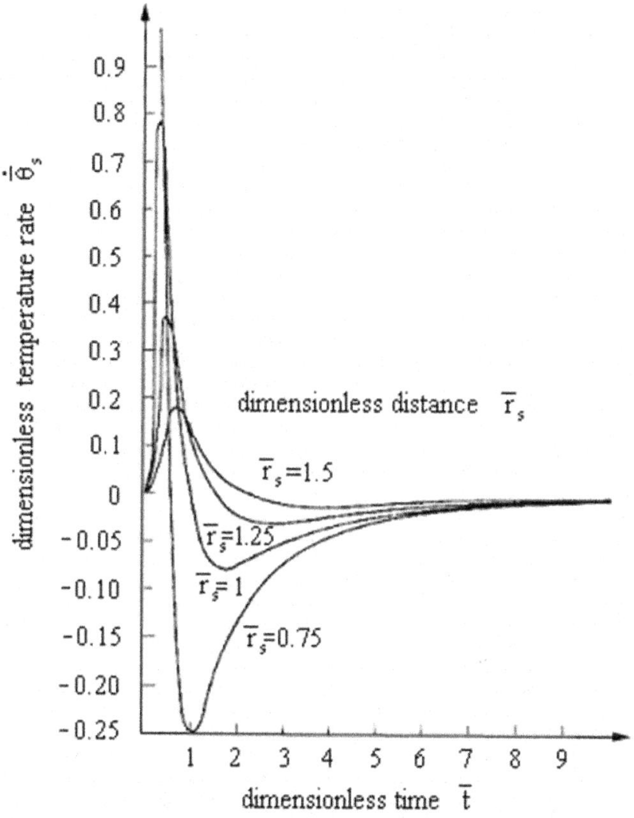

Figure 7-3. Dimensionless cooling rate as a function of dimensionless time for chosen constant dimensionless distances

Fig. 7-3 describes dimensionless cooling rate as a function of dimensionless time for chosen constant dimensionless distances. Fig. 7-4 shows dimensionless cooling rate as a function of dimensionless temperature for chosen constant dimensionless distances.

7.3 Temperature Distribution in a Thin Plate

Consider the heat flow equation (7.2.1) in Cartesian coordinate system (x_1, x_2, x_3).

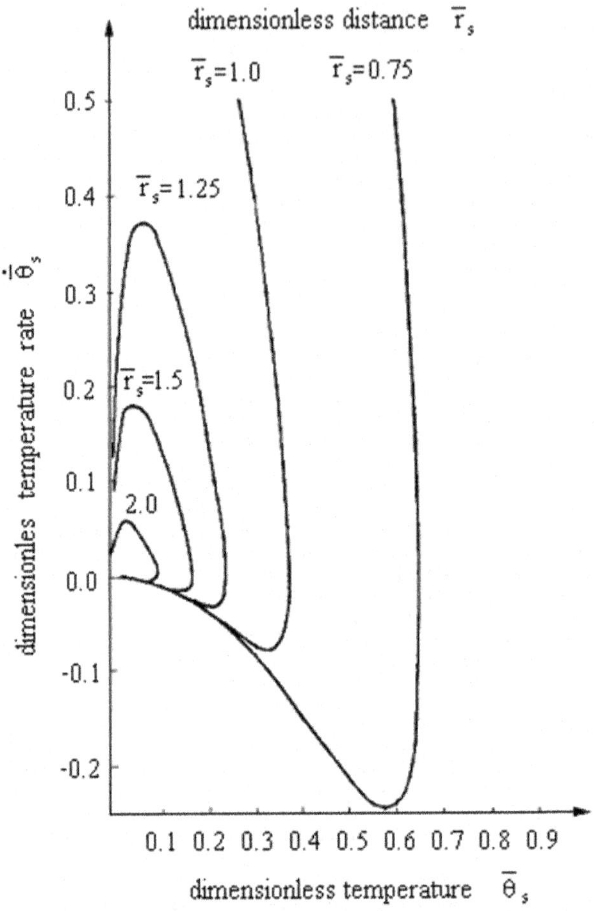

Figure 7-4. Dimensionless cooling rate as a function of dimensionless temperature for chosen constant dimensionless distances

Assume that the heat source (i.e. a welding arc) is moving along the x_1 axis. Define a new coordinate ξ by

$$\xi = x_1 - vt \tag{7.3.1}$$

where v is the constant velocity with which the heat source is moving and ξ is the distance from the point being considered to the heat source along the x_1 axis. Consider the temperature as a function of ξ, x_2, x_3 and t

$$\theta = \theta(\xi, x_2, x_3, t) \tag{7.3.2}$$

We have

$$\frac{d\theta}{dt} = \frac{\partial\theta}{\partial t} + \frac{\partial\theta}{\partial\xi}\frac{\partial\xi}{\partial t} \tag{7.3.3}$$

By Eq. (7.3.1)

$$\frac{d\theta}{dt} = \frac{\partial\theta}{\partial t} - v\frac{\partial\theta}{\partial\xi} \tag{7.3.4}$$

By Eq. (7.3.1) the term $\partial\xi/\partial x_1$ is equal to unity and $\partial^2\theta/\partial x_1^2 = \partial^2\theta/\partial\xi^2$. Finally we get

$$\frac{k}{\rho c}\left(\frac{\partial^2\theta}{\partial\xi^2} + \frac{\partial^2\theta}{\partial x_2^2} + \frac{\partial^2\theta}{\partial x_3^2}\right) = \frac{\partial\theta}{\partial t} - v\frac{\partial\theta}{\partial\xi} \tag{7.3.5}$$

or

$$\alpha\left(\frac{\partial^2\theta}{\partial\xi^2} + \frac{\partial^2\theta}{\partial x_2^2} + \frac{\partial^2\theta}{\partial x_3^2}\right) = \frac{\partial\theta}{\partial t} - v\frac{\partial\theta}{\partial\xi} \tag{7.3.6}$$

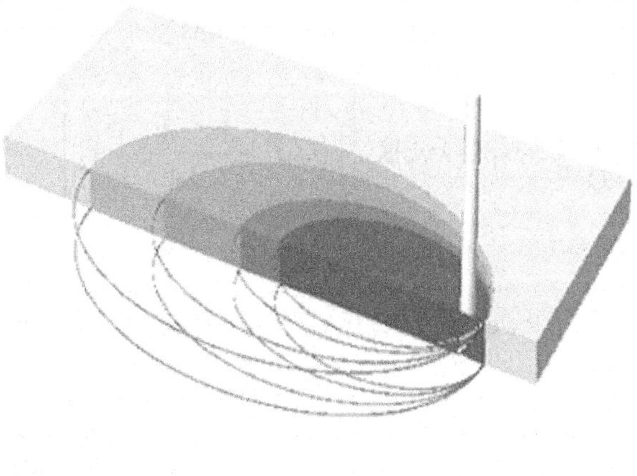

Figure 7-5. Temperature distribution in a thin plate for two dimensional case

Introducing the parameter $2\lambda = 1/\alpha$ we have

$$\frac{\partial^2\theta}{\partial\xi^2}+\frac{\partial^2\theta}{\partial x_2^2}+\frac{\partial^2\theta}{\partial x_3^2}=2\lambda\frac{\partial\theta}{\partial t}-2\lambda v\frac{\partial\theta}{\partial\xi} \tag{7.3.7}$$

If we assume $\partial\theta/\partial t = 0$, then Eq. (7.3.7) takes the form

$$\frac{\partial^2\theta}{\partial\xi^2}+\frac{\partial^2\theta}{\partial x_2^2}+\frac{\partial^2\theta}{\partial x_3^2}=-2\lambda v\frac{\partial\theta}{\partial\xi} \tag{7.3.8}$$

The idealization of temperature distribution in a thin plate is illustrated in Fig. 7-5. It is related to two dimensional case and moving line source. The plate is considered infinite in the ξ and x_2 direction but quite thin and of thickness g in the x_3 direction. The source of heat is considered to be a line source of length g. Assume that there is no temperature gradient in the x_3 direction. Under these assumptions the lines of constant temperature are the same top and bottom. Define the new variables $r_2=\sqrt{\xi^2+x_2^2}$ and $\vartheta = \operatorname{arctg}\xi/x_2$. Consider the boundary conditions

$$\begin{aligned}
&\frac{\partial}{\partial r_2}\theta(r_2,\vartheta,x_3,t)\Big|_{r_2=\infty}=0\\
&-2\pi r_2 k\frac{\partial}{\partial r_2}\theta(r_2,\vartheta,x_3,t)\Big|_{r_2=0}=-2\pi r_2 k\frac{\partial}{\partial r_2}\theta(0,\vartheta,x_3,t)=\frac{\dot{Q}}{g}
\end{aligned} \tag{7.3.9}$$

Assume moreover $\partial\theta/\partial x_3 = 0$ and that heat losses are proportional to $h\theta$. The solution of Eq. (7.3.8) in dimensionless form is

$$\frac{2\pi k g\theta}{\dot{Q}}=\exp(-\lambda v\xi)K_0\left[\left((\lambda v)^2+\frac{h_1+h_2}{kg}\right)^{\frac{1}{2}}r_2\right] \tag{7.3.10}$$

$$\overline{\theta}_2=\exp(-\xi)K_0(\overline{r}_2\mu) \tag{7.3.11}$$

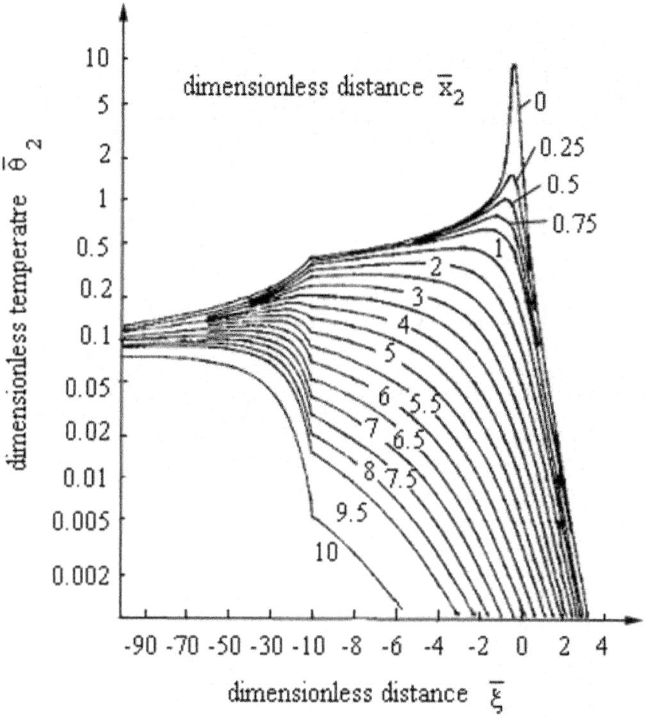

Figure 7-6. Dimensionless temperature as a function of dimensionless distance for chosen parameters $\bar{x}_2$

where

$$\begin{aligned}
&\bar{\theta}_2 = 2\pi kg\theta/\dot{Q} \\
&\bar{\xi} = \lambda v\xi \\
&\mu = \left[1+(h_1+h_2)/(kgv^2\lambda^2)\right]^{\frac{1}{2}} \\
&\bar{r}_2 = \lambda v r_2
\end{aligned} \tag{7.3.12}$$

and h_1 and h_2 are the heat convection coefficient at the top and bottom of the plate and K_0 is the modified Bessel function of the second kind and zero order.

Fig. 7-6 illustrates dimensionless temperature as a function of dimensionless distances for chosen parameters $\bar{x}_2 = \lambda v x_2$. Fig 7-6 was used to construct Fig. 7-7 which shows isotherms for two-dimensional heat flow.

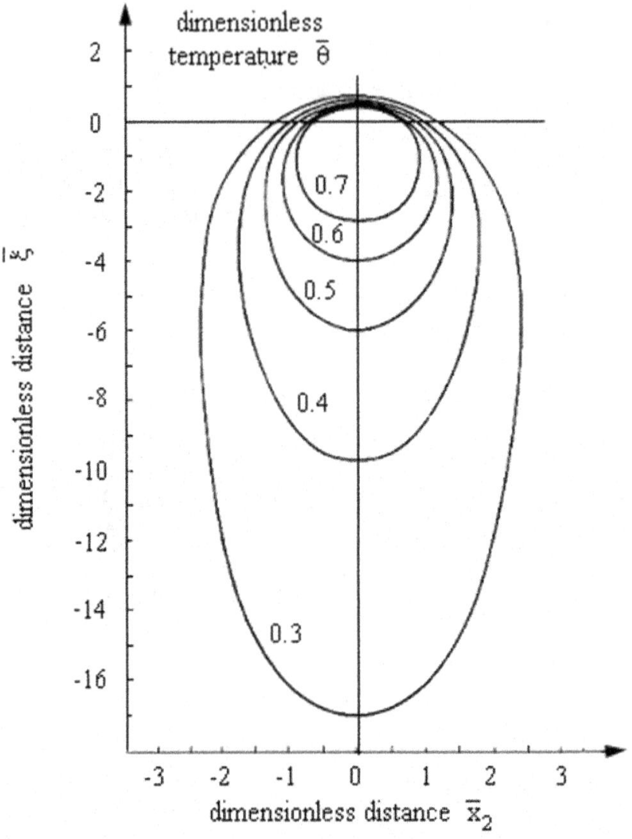

Figure 7-7. Dimensionless temperatures for two dimensional heat flow

Fig. 7-8 shows dimensionless cooling rate as a function of dimensionless distance for chosen parameters $\bar{x}_2$.
In order to obtain cooling rates (Fig. 7-9), we need to differentiate of Eq. (7.3.10) with respect to time. Then

$$\frac{\partial \bar{\theta}_2}{\partial\left(\lambda v^2 t\right)} = \exp\left(-\bar{\xi}\right)\left[K_0\left(\mu \bar{r}_2\right) + \frac{\mu \bar{\xi}}{\bar{r}_2} k_1\left(\mu \bar{r}_2\right)\right] \tag{7.3.13}$$

where K_1 is a modified Bessel function of the second kind and first order

$$\frac{d}{dx} K_0(x) = -K_1(x) \tag{7.3.14}$$

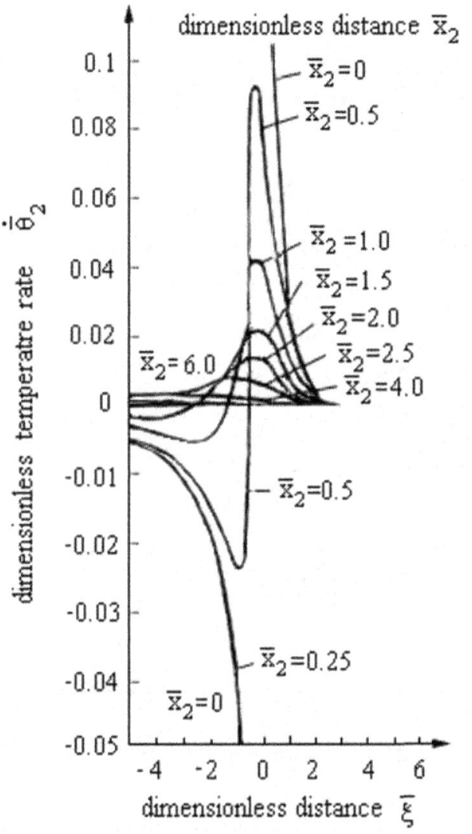

Figure 7-8. Dimensionless cooling rate as a function of dimensionless distance for chosen constant parameter $\bar{x}_2$

The Bessel function K_0 may be approximated by

$$K_0(\mu\bar{r}_2) \cong \exp(-\mu\bar{r}_2)\sqrt{\frac{\pi}{2\pi\bar{r}_2}} \tag{7.3.15}$$

where $\mu\bar{r}_2 > 10$ and the Bessel function K_1 by

$$K_1(\mu\bar{r}_2) \cong \exp(-\mu\bar{r}_2)\sqrt{\frac{\pi}{2\pi\bar{r}_2}} \tag{7.3.16}$$

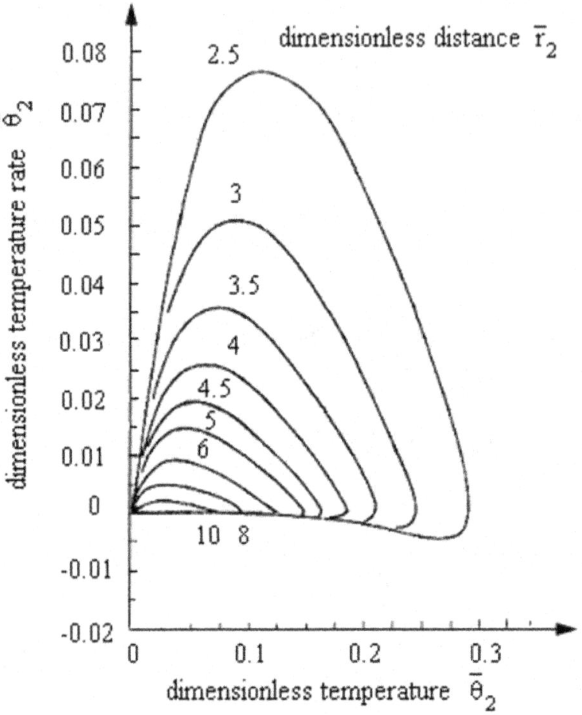

Figure 7-9. Dimensionless cooling rate as a function of dimensionless temperature for chosen values of $\bar{r}_2$

7.4 Temperature Distributions in an Infinitely Thick Plate

The schematic picture for temperature distribution in an infinitely thick plate is shown in Fig. 7-10.

Assume the point heat source and define the variables

$$\begin{aligned} r_3 &= \sqrt{\xi^2 + y^2 + z^2} \\ \bar{r}_3 &= \lambda v r_3 \end{aligned} \tag{7.4.1}$$

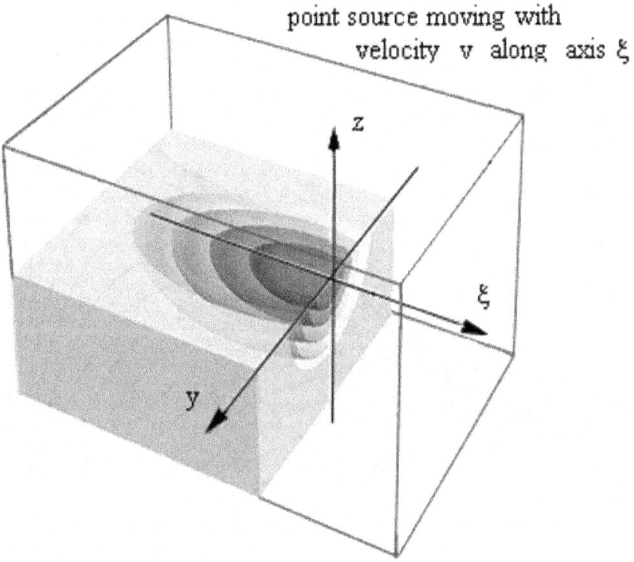

Figure 7-10. Temperature distribution in an infinitely thick plate for three dimensional case

The boundary conditions are

$$\frac{\partial}{\partial r_3}\theta\left(r_{3,}\cdot\right)\Big|_{r_3=\infty}=0 \tag{7.4.2}$$

The solution of Eq. (7.3.8) with the above boundary condition is

$$\frac{4\pi k\theta_3}{\lambda v\dot{Q}}=\frac{\exp\left(-\lambda v r_3-\lambda v\xi\right)}{\lambda v r_3}=\frac{\exp\left(-\bar{r}_3-\xi\right)}{\bar{r}_3} \tag{7.4.3}$$

where the temperature θ with subscript 3 refers to three dimensional case.

Fig. 7-11 is a plot of Eq. (7.4.3) where $\bar{r}_{23}=\lambda v\sqrt{x_2^2+x_3^2}$ and $\bar{\theta}_3=\dfrac{4\pi k\theta_3}{\lambda v\dot{Q}}$. The temperature rate $\dot{\theta}_3=\partial\theta_3/\partial t$ is obtained by differentiating of Eq. (7.4.3) with respect to time to give

$$\frac{4\pi k\dot{\theta}_3}{\dot{Q}\lambda^2v^3}=\frac{\exp\left(-\lambda v r_3-\lambda v\xi\right)}{\lambda v r_3}\left(1+\frac{\xi}{r_3}+\frac{\xi}{\lambda v r_3^2}\right) \tag{7.4.4}$$

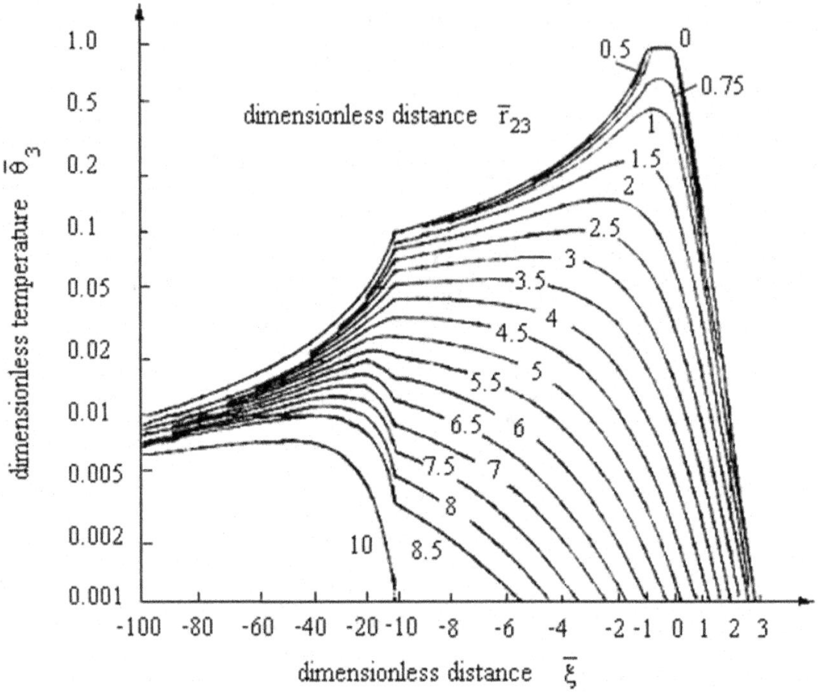

Figure 7-11. Dimensionless temperature as a function of dimensionless distance for chosen parameters $\bar{r}_{23}$

Figure 7-12 is a plot of Eq. (7.4.4) where $\dot{\bar{\theta}}_3 = \frac{4\pi k\dot{\theta}_3}{\dot{Q}\lambda^2 v^2}$. Figure 7-13 is a cross plot of Fig. 7-12 to give temperature rates as a function of temperature.

7.5 Heat Flow in Friction Welding

7.5.1. Thermal Effects in Friction Welding

In friction welding process the heat is produced by direct conversion of mechanical energy to thermal energy at the interface of the workpieces, without the application of electrical energy, or heat from other sources to the workpieces. Friction welds are made by holding a non-rotating workpiece

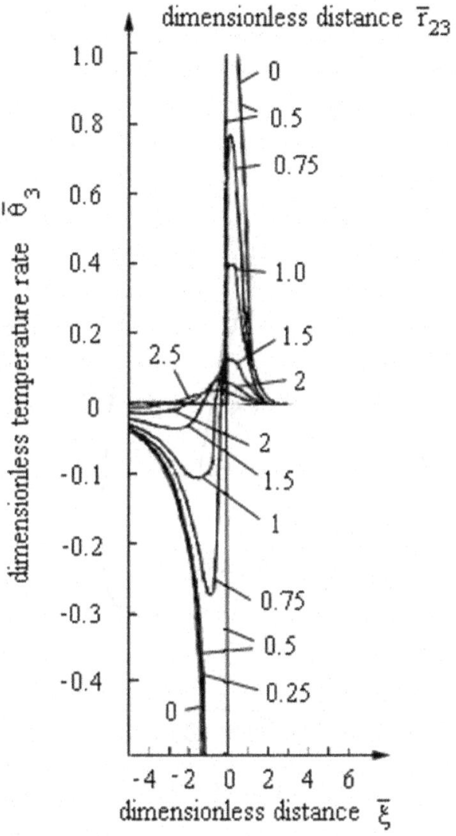

Figure 7-12. Dimensionless cooling rate as a function of dimensionless temperature for chosen parameters $\bar{r}_{23}$

in contact with a rotating workpiece under constant or gradually increasing pressure until the interface reaches welding temperature, and then stopping rotation to complete the weld.

The frictional heat developed at the interface rapidly raises the temperature of the workpieces, over a very short axial distance, to a value approaching, but below, the melting range; welding occurs under the influence of a pressure that is applied while the heated zone is in the plastic temperature range. Friction welding is classified as a solid-state welding process in which joining occurs at a temperature below the melting point of the work metal. If incipient melting does occur, there is no evidence in the finished weld because the metal is worked during the welding stage. In this point we describe some methods for the analysis of the transient temperature

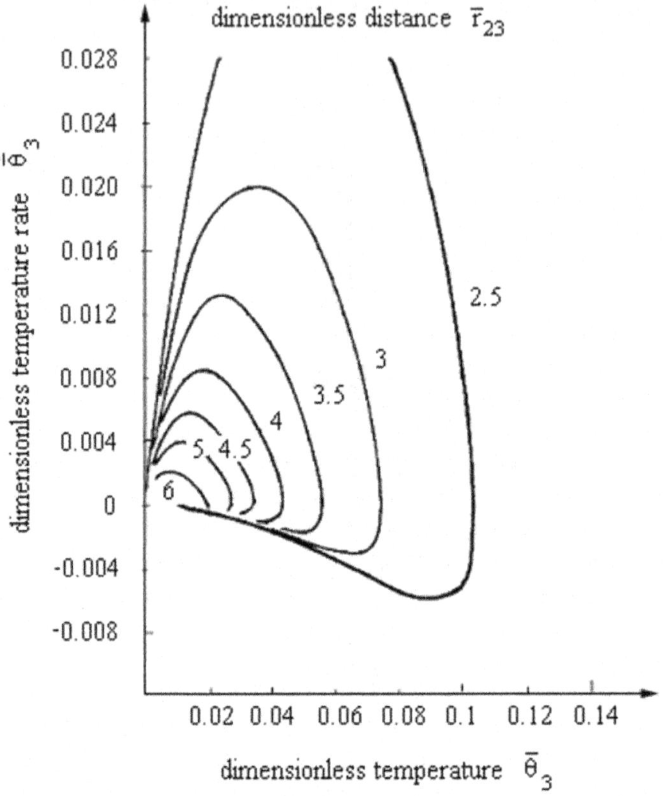

Figure 7-13. Dimensionless cooling rate as a function of dimensionless temperature for chosen parameter $\bar{r}_{23}$

distribution in the vicinity of the weld for arbitrary axisymmetric rods. The common assumption in attempting to provide an analytical solution to such a problem is the postulated temperature independence of all material properties.

7.5.2 Analytical Solutions

Denote the rate of heat generated in the process of friction welding at the place of abutment by q and diameter of the rod by $2r_0$. The heat rate q is usually a function of time t and can be expressed in the form $q(t) = q_0 \exp mt$, where m is the parameter of the process. In some cases the magnitude $q_2 = q/A$ is introduced to characterize the heat rate per unit area where $A = \pi r_0^2$. Consider the heat flow equation in the form

$$\frac{\partial\theta}{\partial t} = \alpha\frac{\partial^2\theta}{\partial x^2} - b\theta + \frac{q_0}{C_\Delta}\exp mt\delta(x) \tag{7.5.1}$$

where x is the coordinate, δ(x) is the Dirac function, $b = 2h/C_\Delta r_0$ and h is the surface film conductance.
The boundary conditions are

$$\frac{\partial}{\partial x}\theta\,(\pm\infty, t) = 0 \qquad \theta\,(x,0) = 0 \tag{7.5.2}$$

Using a Laplace transformation

$$\tilde{\theta}\,(x,s) = \int_0^\infty \theta\,(x,t)e^{-st}dt \tag{7.5.3}$$

we get

$$\alpha\tilde{\theta}''(x,s) - (b+s)\tilde{\theta}(x,s) + \frac{q_0}{C_\Delta(s+m)} = 0 \tag{7.5.4}$$

$$\tilde{\theta}'(\pm\infty, s) = 0 \tag{7.5.5}$$

Since

$$\tilde{\theta}(-x,s) = \tilde{\theta}(x,s) \tag{7.5.6}$$

we have

$$\tilde{\theta}(x,s) = \frac{q_0\sqrt{\alpha}}{2kA(s-m)\sqrt{(s+b)}} \times \exp\left(-\sqrt{(s+b)}\cdot\frac{x}{\sqrt{\alpha}}\right) \quad x \geq 0 \tag{7.5.7}$$

Hence we obtain

$$\theta(x,t)=\frac{q_0\sqrt{\alpha}}{2kA}\cdot\frac{\exp mt}{2\sqrt{(b+m)}}\left\{\exp\left[-\sqrt{\left(\frac{b+m}{\alpha}\right)}\cdot x\right]\right.$$
$$\times\operatorname{erfc}\left[\frac{x}{2\sqrt{(\alpha t)}}\sqrt{((b+m)t)}\right] \tag{7.5.8}$$
$$\left.-\exp\left[\sqrt{\left(\frac{b+m}{\alpha}\right)}\cdot x\right]\operatorname{erfc}\left[\frac{x}{2\sqrt{(\alpha t)}}+\sqrt{((b+m)t)}\right]\right\}$$

Introduce the new function

$$\psi_1(\rho_1,\tau)=\frac{1}{2}\operatorname{erfc}\left(\frac{\rho_1}{2\sqrt{\tau}}-\sqrt{\tau}\right)$$
$$-\frac{1}{2}\exp^2\rho_1\operatorname{erfc}\left(\frac{\rho_1}{2\sqrt{\tau}}+\sqrt{\tau}\right) \tag{7.5.9}$$

where

$$\rho_1=\sqrt{\left(\frac{b+m}{\alpha}\right)}x \qquad \tau=(m+b)t \tag{7.5.10}$$

Then we get from equation (7.5.8)

$$\theta(x,t)=\frac{q_0}{2kA}\sqrt{\left(\frac{\alpha}{b+m}\right)}\exp\left[mt-x\sqrt{\left(\frac{b+m}{A}\right)}\right]$$
$$\times\psi_1\left[\sqrt{\left(\frac{b+m}{\alpha}\right)}\cdot x,(b+m)t\right] \tag{7.5.11}$$

For m = 0, b > 0 and then Eq. (7.5.11) has the form

$$\theta(x,t) = \frac{q_0}{2kA}\sqrt{\left(\frac{\alpha}{b}\right)} \exp\left[-x\sqrt{\left(\frac{b}{\alpha}\right)}\right] \times \psi_1\left[\sqrt{\left(\frac{b}{\alpha}\right)} \cdot x, bt\right] \quad x \geq 0 \tag{7.5.12}$$

For small values of the x Eq. (7.5.8) can be expressed with sufficient accuracy in the following form

$$\theta(x,t) \cong \frac{q_0}{2kA} \exp mt\left[\sqrt{\left(\frac{\alpha}{b+m}\right)} \operatorname{erf}\sqrt{((b+m)t)} - x\right] \quad x \geq 0 \tag{7.5.13}$$

Consider the case (b + m) = 0. Then Eq. (7.5.8) has the form

$$\theta(x,t) = \frac{q_0}{kA} \exp(-bt)\sqrt{(\alpha t)} \times \operatorname{ierfc}\frac{x}{2\sqrt{(\alpha t)}} \quad x \geq 0 \tag{7.5.14}$$

where

$$\operatorname{ierfc} u = \int_u^\infty \operatorname{erfc} u \, du \quad \operatorname{erfc} u = \frac{2}{\sqrt{\pi}}\int_u^\infty e^{-x^2} dx \quad \operatorname{erfc}(\infty) = 0 \tag{7.5.15}$$

At the place of contact when $bt << 1$ the temperature can be expressed as

$$\theta(0,t) = \frac{q_0}{\sqrt{(\pi kA)}} \exp(-bt)\sqrt{(\alpha t)} \tag{7.5.16}$$

If $m = b = 0$ and q_0 = const, then the temperature in the rods at the place of abutment can be described by the equation

$$\theta(0,t) = \frac{q_0\sqrt{t}}{\sqrt{(\pi k C_\Delta A)}} \tag{7.5.17}$$

The case b + m < 0 appears in the process of cooling. Denote the time of heating by t_c. For $t \geq t_c$ we get

$$\theta(x,t) = \frac{q_0}{2kA}\sqrt{\left(\frac{\alpha}{b+m}\right)} \exp\left(mt - x\sqrt{\left(\frac{b+m}{\alpha}\right)}\right)$$
$$\times\left\{\psi_1\left[\sqrt{\left(\frac{b+m}{\alpha}\right)}\cdot x, (b+m)t\right]\right. \tag{7.5.18}$$
$$\left. - \psi_1\left[\sqrt{\left(\frac{b+m}{\alpha}\right)}\cdot x, (b+m)(t-t_c)\right]\right\}$$

For x = 0 the temperature is given by

$$\theta(0,t) = \frac{q_0}{2kA}\sqrt{\left(\frac{\alpha}{b+m}\right)} \tag{7.5.19}$$
$$\times \exp mt\left[\operatorname{erf}\sqrt{((b+m)t)} - \operatorname{erf}\sqrt{((b+m)(t-t_c))}\right]$$

Consider the variation of temperature in rods along the radius in cylindrical coordinate system (r, x). The heat flow equation in this case can be written as

$$C_\Delta \frac{\partial\theta}{\partial t} = k\left(\frac{\partial^2\theta}{\partial r^2} + \frac{1}{r}\frac{\partial\theta}{\partial r} + \frac{\partial^2\theta}{\partial x^2}\right) \tag{7.5.20}$$
$$+ [q_2(r) - q_2]\delta(x)$$

The boundary conditions have the form

$$\frac{\partial}{\partial x}\theta(r, \pm\infty, t) = 0 \qquad \frac{\partial}{\partial r}\theta(r_0, x, t) = 0$$
$$\frac{\partial}{\partial r}\theta(0, x, t) = 0 \qquad \theta(r, x, 0) = 0 \tag{7.5.21}$$

Introduce the following dimensionless magnitudes

$$\bar{\theta} = \frac{k\theta}{r_0 q_2} \tag{7.5.22}$$

$$\bar{t} = \frac{\alpha t}{r_0^2} \tag{7.5.23}$$

$$\bar{r} = r / r_0 \qquad \xi = x / r_0 \tag{7.5.24}$$

$$f(r) = \frac{q_2(r)}{q_2} - 1 \tag{7.5.25}$$

Then the equation for dimensionless temperature $\bar{\theta}(\bar{r},\xi,\bar{t})$ has the form

$$\begin{aligned}\bar{\theta}(\bar{r},\xi,\bar{t}) &= \int_0^{\bar{t}} \frac{d\bar{t}'}{\sqrt{4\pi(\bar{t}-\bar{t}')}} \exp\left[-\frac{\xi^2}{4(\bar{t}-\bar{t}')}\right] \\ &\times 2\sum_{n=0}^{\infty} \exp\left[-\mu_n^2(\bar{t}-\bar{t}')\right]\frac{F(\mu_n)}{J_0^2(\mu_n)} J_0(\mu_n \bar{r})\end{aligned} \tag{7.5.26}$$

where the heat source is given as $f(\bar{r}) = \delta(\bar{t}-\bar{t}')$, $\bar{t}'$ is the heating time, μ_n are solutions of the equation $J_0'(\mu_n) = J_1(\mu_n) = 0$, $n = 1, 2, \ldots$;
$F(\mu_n) = \int_0^1 f(\bar{r}) J_0(\mu_n \bar{r}) \bar{r}\, d\bar{r}$ and J_0, J_1 are Bessel functions.

Equation (7.5.26) can be written as

$$\begin{aligned}\bar{\theta}(\bar{r},\xi,\bar{t}) &= \sum_{n=0}^{\infty} \frac{F(\mu_n)}{J_0^2(\mu_n)} J_0(\mu_n \bar{r}) \\ &\times \int_0^{\tau} \frac{d\bar{t}}{\sqrt{(\pi t)}} \exp\left(-\mu_n^2 \bar{t} - \frac{\xi^2}{4\bar{t}}\right)\end{aligned} \tag{7.5.27}$$

or introducing a function $\psi_1(\bar{r},\bar{t})$

$$\begin{aligned}\bar{\theta}(\bar{r},\xi,\bar{t}) &= \sum_{n=0}^{\infty} \frac{F(\mu_n)}{\mu_n J_0^2(\mu_n)} J_0(\mu_n \bar{r}) \\ &\times \exp(-\mu_n \bar{r}) \psi_1(\mu_n \xi, \mu_n^2 \bar{t})\end{aligned} \tag{7.5.28}$$

For $\xi = 0$ we have

$$\bar{\theta}(\bar{r}, 0, \bar{t}) = \sum_{n=0}^{\infty} \frac{F(\mu_n)}{\mu_n J_0^2(\mu_n)} J_0(\mu_n \bar{r}) \operatorname{erf} \mu_n \sqrt{\bar{t}} \tag{7.5.29}$$

For $\bar{r} = 0$ and $\bar{r} = 1$ we have respectively

$$\bar{\theta}(0, 0, \bar{t}) = \sum_{n=0}^{\infty} \frac{F(\mu_n)}{\mu_n J_0^2(\mu_n)} \operatorname{erf} \mu_n \sqrt{\bar{t}} \tag{7.5.30}$$

$$\bar{\theta}(1, 0, \bar{t}) = \sum_{n=0}^{\infty} \frac{F(\mu_n)}{\mu_n J_0(\mu_n)} \operatorname{erf} \mu_n \sqrt{\bar{t}} \tag{7.5.31}$$

In order to get a solution for $\bar{\theta}(\bar{r}, \xi, \bar{t})$ it is necessary to define q_2 as a function of radius $\bar{r}$.
Assume

$$q_2(\bar{r}) = q_2 \sum_{m=1}^{\infty} \varepsilon_m \bar{r}^m \tag{7.5.32}$$

Then we can write

$$\begin{aligned} F(\mu) &= \int_0^1 \left[\sum_{m=1}^{\infty} \varepsilon_m \bar{r}^m - 1 \right] J_0(\mu \bar{r}) \bar{r} \, d\bar{r} \\ &= \sum_{m=1}^{\infty} \varepsilon_m \int_0^1 J_0(\mu \bar{r}) \bar{r}^{m+1} d\bar{r} \\ &= \sum_{m=1}^{\infty} \varepsilon_m F_m(\mu) \end{aligned} \tag{7.5.33}$$

where

$$F_m(\mu) = \int_0^1 J_0(\mu\bar{r})\,\bar{r}^{m+1}\,d\bar{r}$$

$$= \frac{1}{\mu}J_1(\mu) + \frac{m}{\mu^2}J_0(\mu) - \frac{m^2}{\mu^2}F_{m-2}(\mu) \qquad (7.5.34)$$

$$m \geq 0$$

$$F_0(\mu) = 0$$

$$F_1(\mu) = \frac{1}{\mu^2}J_0(\mu) - \frac{1}{\mu^2}\int_0^1 J_0(\mu\bar{r})\,d\bar{r} \qquad (7.5.35)$$

$$= J_0(\mu)H_1(\mu) - J_0(\mu)H_0(\mu)$$

and $H_0(\mu)$, $H_1(\mu)$ are the so-called Struve functions.

Chapter 8

NUMERICAL SOLUTIONS OF THERMAL PROBLEMS IN WELDING

8.1 Introduction

The effective solutions of complex thermal problems in welding only recently has become possible. In the last two decades one observes the vigorous development of effective numerical methods of analysis. Automobile, aircraft, nuclear and ship industry are experiencing a rapidly-growing need for the numerical tools to handle complex problems of welding. Efficient numerical methods for nonlinear welding problems are needed because experimental testing in such cases is often prohibitively expensive or physically impossible.

The intention of this chapter is to discuss numerical solutions of heat flow in chosen welding processes. The choice of topics is by no means comprehensive and is based on the authors results concerning laser and electroslag welding. In this chapter heat flow problems in arc-welding and friction welding are omitted to do not repeat the results which are in agreement with analytical solutions of there problems presented in Chapter 7.

8.2 Temperature Distributions in Laser Microwelding

Laser welding has a wide application in industry. The pulsed lasers allow the production of impulses of radiation in the time intervals $10^{-3} - 10^{-12}$ sec, obtaining the peak power in the range of gigawatts and radiation intensity to 10^{17} W/m^2. It is an important problem in laser welding to determine both temperature field within welded material and changes of material's shape during melting process for a given laser impulse.

Optical properties of metals are implied directly from their electrical properties. An electric conductivity, which is dependent on material properties, its temperature and frequency ω of falling electro-magnetic wave, has important significance. The plasma frequency ω_p for metals implying from quantity of electrons of conduction per unit volume ($N = 10^{28} - 10^{29}$ elec/m^3) is equal to $10^{15} - 10^{16}$ /sec. For wave frequency $\omega < \omega_p$ the reflection coefficient is very high and a part of absorbed wave is strongly dumped. Its intensity decreases in an exponential way. For waves of frequency $\omega > \omega_p$ metals are to be transparent, and for instance for $\omega_p = 5.7 \times 10^{15}$ /sec a material becomes transparent for waves shorter than 330 nm (i.e. in ultraviolet). In the range of visible infrared radiation for almost all metals the reflection coefficient is high and reaches values up to 90% and more. However, it is strongly dependent on wave frequency and, for instance, for silver, in the visible range is equal to ~ 95%, and in ultraviolet 4.2%. The value of reflection coefficient depends in a significant way on the kind of machining and the cleanliness of the material surface.

For temperatures in which the electric super-conductivity appears, optical properties change because of abnormal skin effect and need be considered in another way. Using a simplified model of monochromatic radiation absorption (only by free electrons) on the surface of a semi-infinite body we may assume that in the visual and infrared parts of the radiation spectrum the absorption is in the layer of thickness $\delta < 0.1$ μm.

In the absorption model the value of the absorption coefficient for metals is equal to $10^8 - 10^7$ [1/m], and within the body by the Bouguer law the heat source is

$$A(z,t) = E_a(t)\beta e^{-\beta z} \tag{8.2.1}$$

where E_a is absorbed, changed in time radiation intensity of the source, β is the absorption coefficient for given material, t is the time variable and z is the coordinate.

Assuming Eq. (8.2.1) we may easily determine the thickness of the layer in which fixed part of radiation will be absorbed, The results of calculations for extremal values of β are given in Table 8-1. Experimental data confirm that visual and infrared radiation is absorbed in metal within the layer of thickness 0.01 – 0.1 μm. In a large number of papers the model of laser radiation absorption on surface of body is analyzed. Assuming such a model we should analyze the time intervals for which this model is correct, because, for instance, the laser impulses of duration $10^{-12} - 10^{-3}$ sec cannot be treated in the same way. For laser impulses of duration 10^{-8} sec and longer we can analyze the temperature field in classical sense, because for these times we have a balance between the electrons' and mesh's temperature (relaxation time for metals is equal to $10^{-11} - 10^{-13}$ sec).

Table 8-1. Absorption coefficient β

Absorption coefficient $\beta\left[\frac{1}{m}\right]$	Thickness of the layer in which fixed part of radiation is absorbed 99%	99,9%
10^8	$z = 4.605 \times 10^{-8}$ m	6.907×10^{-8} m
10^7	$z = 4.605 \times 10^{-7}$ m	6.907×10^{-7} m

We should also check whether the diffusion mechanism for shorter times i.e. $10^{-8} - 10^{-9}$ sec is sufficiently strong, in order that the model of surface absorption does not deform the obtained temperature field.

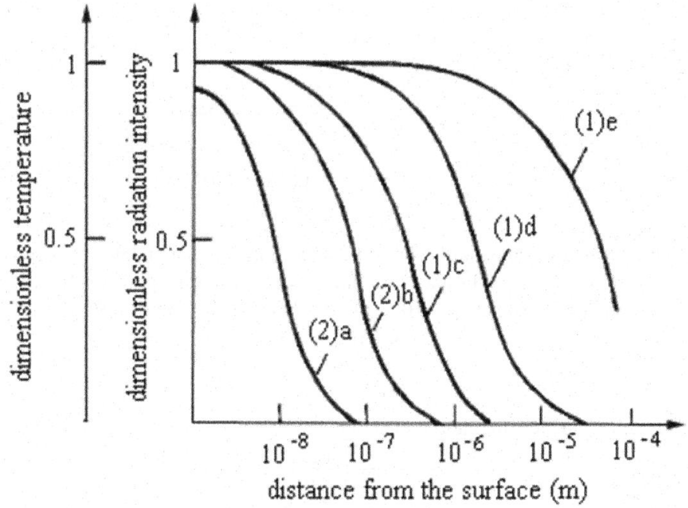

Figure 8-1. Changes of (1) dimensionless temperature and (2) dimensionless radiation intensity E(z)/E within the material as a function of the distance from the surface. (a) absorption coefficient $\beta = 10^8$, (b) absorption coefficient $\beta = 10^7$, (c) $\alpha t = 65 \times 10^{-15}$, (d) $\alpha t = 65 \times 10^{-13}$, (e) $\alpha t = 65 \times 10^{-12}$

To determine the thermal perturbation domain appearing during conduction mechanism and reciprocate it with the region in which the source occurs the simple calculations have been done by solving the one-dimensional heat conduction equation

$$\frac{\partial\theta(z,t)}{\partial t} = \alpha\frac{\partial^2\theta(z,t)}{\partial z^2} \tag{8.2.2}$$

The results for various values of αt are shown in Figs. 8-1 and 8-2. The changes of radiation intensity for interesting values only [$\alpha t \geq 6.5 \times 10^{-14}$ (m^2)] are presented in these figures. [$E(z) = E_a e^{-\beta z}$]. The times for given

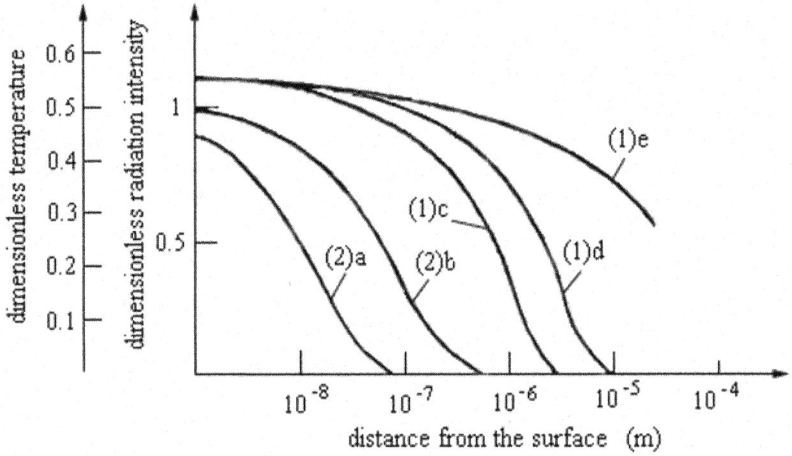

Figure 8-2. Dimensionless temperature (1), and dimensionless radiation intensity (2) within the material as a function of the distance from the surface. (a) absorption coefficient $\beta = 10^8$, (b) absorption coefficient $\beta = 10^7$, (c) $\alpha t = 65 \times 10^{-14}$, (d) $\alpha t = 65 \times 10^{-13}$, (e) $\alpha t = 65 \times 10^{-12}$

Table 8-2. Thermal parameters for chosen materials

Material	Titanium	Wolfram	Copper 99,999%
Thermal diffusivity α (293 K) [m^2/s]	6.36×10^{-6}	65×10^{-6}	116.4×10^{-6}
αt [m^2]: 6.5×10^{-14}	1.02×10^{-8} †	$\times 10^{-9}$ †	0.56×10^{-9} †
αt [m^2]: 6.5×10^{-12}	1.02×10^{-6} †	$\times 10^{-7}$ †	0.56×10^{-7} †

† Time (sec)

parameters αt for different thermal properties of materials are shown in Table 8-2. The graphs show that the domain of thermal perturbation appearing as a result of conduction for $\alpha t = 65 \times 10^{-14}$ is several times greater than the domain in which the volumetric source decreases 100 times (for $\beta = 10^7$ 1/m); for $\alpha t > 65 \times 10^{-13}$ this domain is over one order of magnitude higher. Figures 8-1 and 8-2 indicate that we should use the condition on surface absorption for times longer than 10^{-7} sec and even for good conducting metals longer than 10^{-8} sec.

The changes of radiation intensity during laser impulse may be described sufficiently by the sum of exponential functions or as the product of exponential and power functions

$$\begin{aligned} E(t) &= B_1 \exp(-v_1 t) - B_2 \exp(-v_2 t) \\ E(t) &= B (vt)^n \exp(-v_3 t) \end{aligned} \tag{8.2.3}$$

where constants B, v and n are taken to be dependent on the kind of laser head, peak power and the duration of impulse time. Changing these constants in Eq. (8.2.3) we may also analyze reflectivity of material subjected to laser light and obtain conditions determining the intensity of absorbed radiation. Another problem is to find the distribution of power in the cross-section of the laser beam. For the focal beam of radiation produces on the surface of material the power distribution is described by the Gauss function. For a multimodal beam, the power distribution in the cross-section of the beam depends on construction of an optical resonator and conditions of its work and it may be nonhomogeneous. In modern laser head systems the laser beam has homogeneous power distribution in the cross-section, and in some cases a drop of power appears in its extreme regions.

The models of laser microwelding woll be analyzed using a finite element method. An example is presented in Fig. 8-3. We assume the following boundary conditions on the boundaries of the analyzed region (see Fig. 8-3):

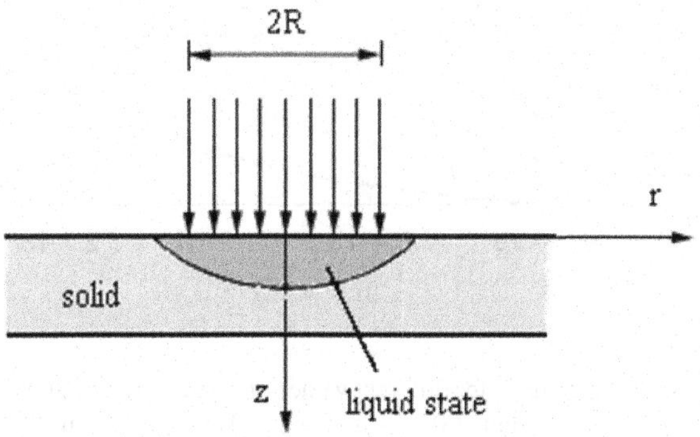

Figure 8-3. Elements subjected to laser microwelding

$r \leq R$

$$-k\frac{\partial \theta}{\partial n}\bigg|_{a_1} = E_q(r,t)\big|_{z=0} + \in(t)\,\sigma_0\left[\theta(r,t)^4\big|_{z=0} - \theta_\infty^4\right]$$
$$-h(t)\left[\theta(r,t)\big|_{z=0} - \theta_\infty\right]$$

$r > R$ (8.2.4)

$$-k\frac{\partial \theta}{\partial n}\bigg|_{a_1} = \in(t)\sigma_0\left[\theta(r,t)^4\big|_{z=0} - \theta_\infty^4\right] - h(t)\left[\theta(r,t)\big|_{z=0} - \theta_\infty\right]$$
$$-k\frac{\partial \theta}{\partial n}\bigg|_{a_2} = h_1(\theta - \theta_\infty)$$

where $\in$ is the coefficient of emissivity, σ_0 is the Stefan-Boltzmann constant, h_1 is the surface film conductance, a_1, a_2 are surfaces on the top and bottom of the element and θ_∞ is the environmental temperature.

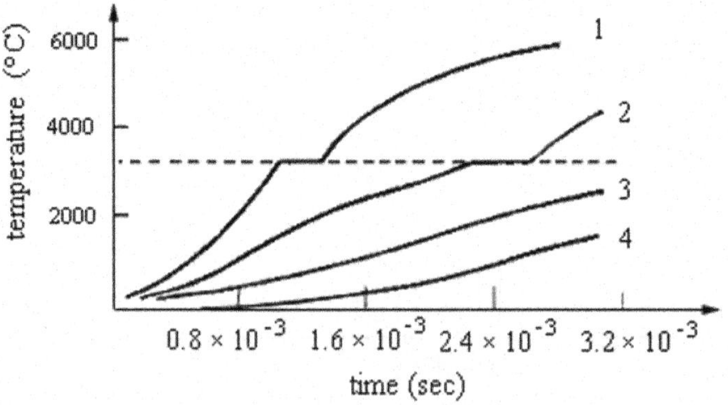

Figure 8-4. Temperatures for different values of z and t. (1) $z = 0.05$ mm, (2) $z = 0.15$ mm, (3) $z = 0.25$ mm, (4) $z = 0.45$ mm

All numerical examples illustrating the microwelding process are analyzed with the assumption that the phase change occurs at a specified temperature. Let us consider the problem of laser beam action when $2R = 3$ mm. The finite element meshes used in the analysis has been omitted because of its low importance. Material properties are given in Table 8-3. Fig. 8-4 shows temperature distribution on the axis (r = 0) for four values of coordinate z. In Fig. 8-5 we present the shape of normal laser impulse heating the element. This impulse is described by the formula

$$E_z(t) = 2.4 \times 10^{10}\left[\exp(-750\,t) - \exp(-100\,t)\right] \tag{8.2.5}$$

where t is seconds. We assume a constant value of radiation intensity at the cross-section of the beam.

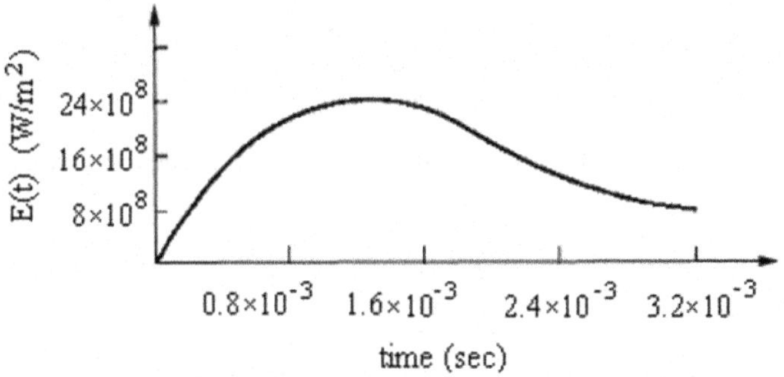

Figure 8-5. Radiation intensity absorbed by metal

Table 8-3. Thermal properties of tungsten and copper

Property	Tungsten		Copper	
Melting temperature [°C]		3280		1083
Latent heat [kJ/m^3]		3.357×10^6		1.902×10^6
Specific heat [kJ/m^{3}K]	$\theta < 2130$	$c = 2476 \times 0.3758\,\theta$	$0 < \theta < 1083$	$c = 1357.36\,\theta - 367157$
	$2130 < \theta < 3280$	$c = 1743 \times 6.68\,\theta$	$\theta > 1083$	$c = 3573$
	$\theta > 3280$	$c = 4200$		
Thermal conductivity [W/mK]	$\theta \leq 3280$	$k = 135.71 - 0.02\,\theta$	$0 < \theta < 1083$	$k = 340$
	$\theta > 3280$	$k = 66.16 - 0.01\,\theta$	$\theta > 1083$	$k = 315$
Viscocity [kg/m·s]		0.013		0.007
Density [kg/m^3]		19.3		8.96
Thermal expansion coefficient $\times 10^6$ K^{-1}		4		16.6

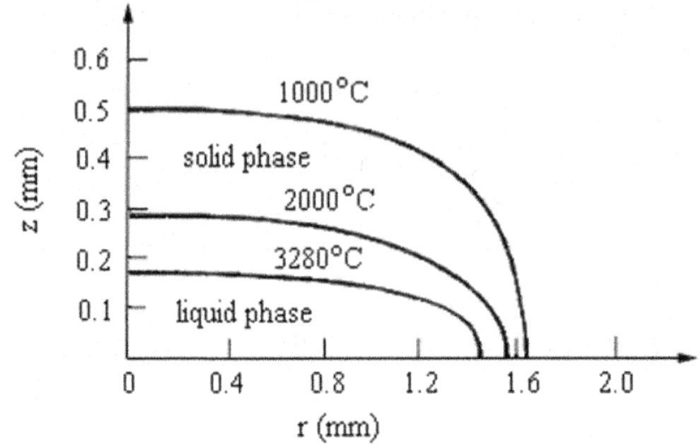

Figure 8-6. Temperature field after 2.76×10^{-3} sec. Laser impulse described by Eq.(8.2.5)

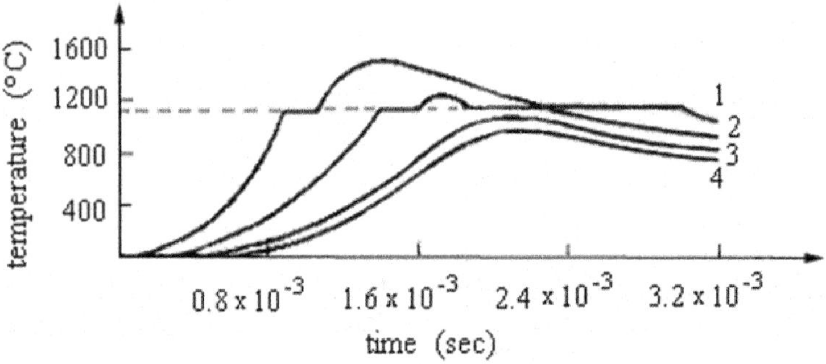

Figure 8-7. Temperature field in copper specimen for points lying on the axis (r = 0); (1) z = 0.05 mm, (2) z = 0.15 mm, (3) z = 0.20 mm, (4) z = 0.30 mm

The maximal value of power density appears for this impulse after approximately 1.2×10^{-3} sec and is equal to 25.3×10^{8} W/m^2. From the obtained results we see that this impulse causes a melting effect. The distribution of temperatures after 2.76×10^{-3} sec is presented in Fig. 8-6. As we can see the metal is fused to a depth of 0.17 mm. In Fig. 8-7 we present the calculated changes of temperature for copper specimen on its axis as a function of four values of z. In Fig. 8-8 we show the graph describing the radiation intensity of a laser beam

$$E_\alpha(t) = 10^9 (3200t)^5 \exp(-5000t) \tag{8.2.6}$$

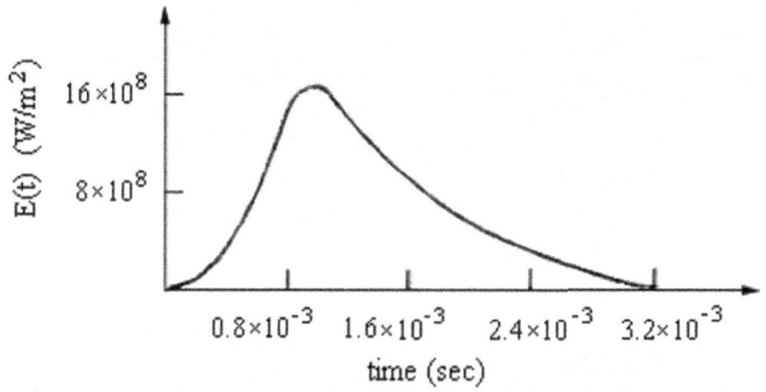

Figure 8-8. Intensity of laser impulse described by Eq. (8.2.6)

The presented results refer to an impulse of constant intensity in the cross-section of the beam. For $z = 0.05$ mm the maximum of temperature is reached for a time of approximately 1.4 msec. The distribution of isotherms in the copper specimen after the time 1.8×10^{-9} sec is presented in Fig. 8-9.

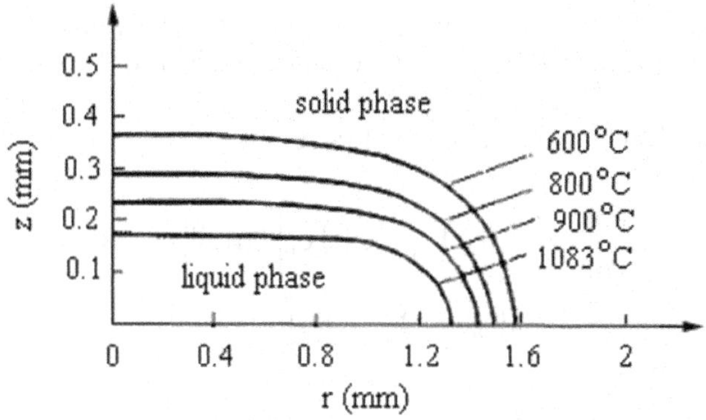

Figure 8-9. Temperature field in copper specimen after 1.8×10^{-3} sec

Based on the calculated temperature fields we may analyze the influence of both thermal properties of materials and shape or parameters of laser impulse on the melting process in laser microwelding. In order to obtain melting of metal during normal impulses it is necessary to apply the power density 10^9– 10^{10} W/m^2. For metals of lower melting temperature we may use the power density in the range of 10^8 W/m^2. For the copper and laser impulse described by Eq. (8.2.5) for $z = 0.05$ mm on the axis of specimen, for the times 0.65 and 0.95 msec the values $\Delta\theta/\Delta t$ change from 1.95×10^6 to

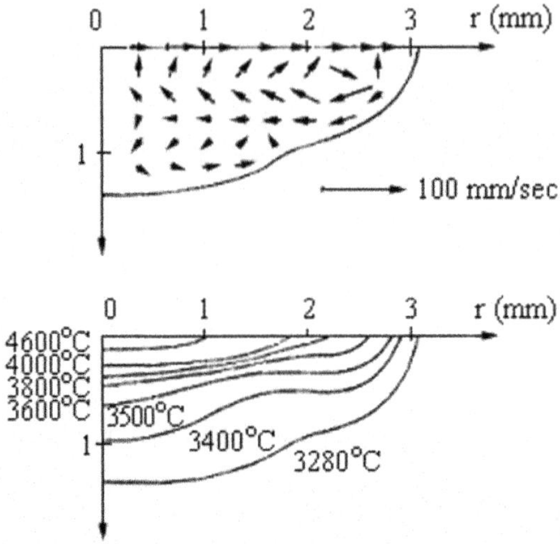

Figure 8-10. Metal flow and temperatures in tungsten specimen

1.66×10^6 K/sec for the solid state. For t = 0.8 msec the values $\Delta\theta/\Delta z$ for $0.05 < z < 0.15$ mm and $0.20 < z < 0.30$ mm are equal to 3.93×10^6 and 1.37×10^6 K/m.

Assume the following boundary conditions

$$r \le R \qquad -k\frac{\partial\theta}{\partial n}\bigg|_{a_1} = E_a(r,t)\big|_{z=0} + \in(t)\,\sigma_0\left[\theta(r,t)^4\big|_{z=0} - \theta_\infty^4\right] - h(t)\left[\theta(r,t)\big|_{z=0} - \theta_\infty\right] \qquad (8.2.7)$$

$$r > R \qquad -k\frac{\partial\theta}{\partial n}\bigg|_{a_1} = \in(t)\,\sigma_0\left[\theta(r,t)^4\big|_{z=0} - \theta_\infty^4\right] - h(t)\left[\theta(r,t)\big|_{z=0} - \theta_\infty\right] \qquad (8.2.8)$$

$$-k\frac{\partial\theta}{\partial n}\bigg|_{a_2} = h_1(\theta - \theta_\infty)$$

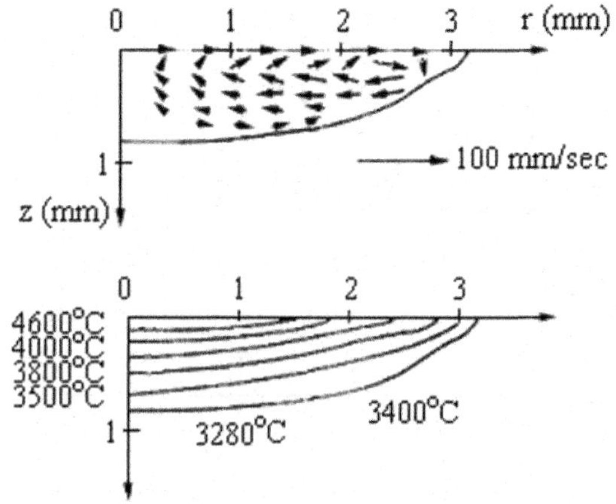

Figure 8-11. Metal flow and temperature field in tungsten specimen

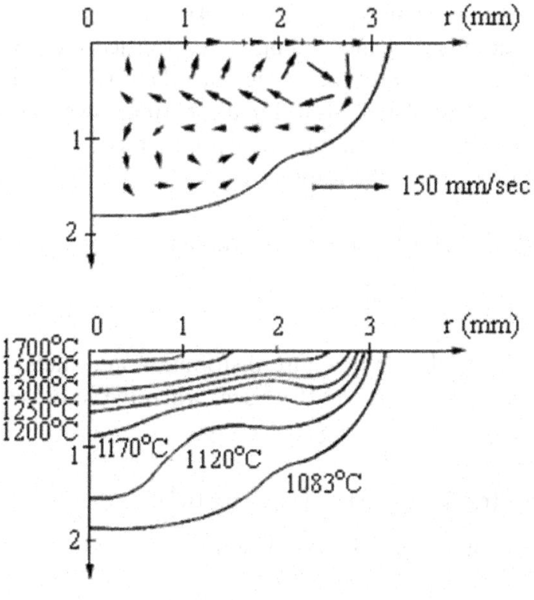

Figure 8-12. Flow of metal and temperatures in cooper specimen

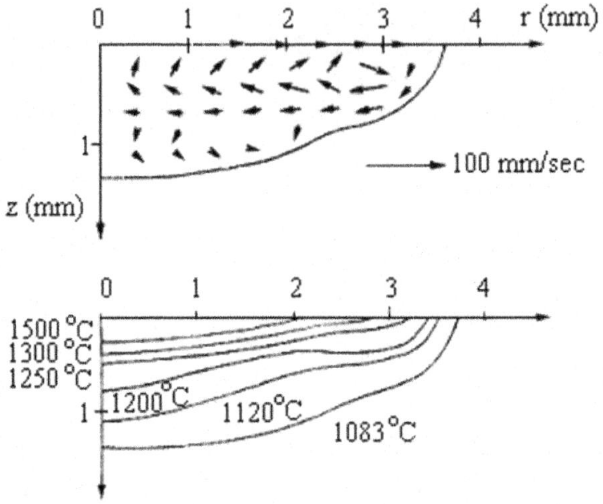

Figure 8-13. Flow of metal and temperatures in copper specimen

with the same notations as in expression (8.2.4).

Assume that the phase change occurs at a specified temperature. Let us consider the problem of laser beam action when R = 3 mm. Material properties are given in Table 8-3. Figure 8-10 shows the temperature distribution and metal flow for a tungsten specimen and time 3×10^{-2} sec. This impulse is described by the formula

$$E_a(r,t) = 0.28 \times 10^{14}[\exp(-7000t) - \exp(-1000t)]\exp(-0.3r^2) \quad (8.2.9)$$

Figure 8-11 shows the temperature distribution and metal flow for the tungsten specimen and time 0.5×10^{-2} sec, when the impulse is described by the formula

$$E_a(r,t) = 0.24 \times 10^{16}\left[\exp(-6 \times 10^3 t) - \exp(-2 \times 10^3 t)\right]\exp(-0.125r^2) \quad (8.2.10)$$

Figure 8-12 shows the temperatures and metal flow for a copper specimen and time 2×10^{-2} sec, when the impulse is given as

$$E_a(r,t) = 0.12 \times 10^{11}(32 \times 10^3 t)\exp(-5 \times 10^4 t)\exp(-0.2r^2) \qquad (8.2.11)$$

and Fig. 8-13 gives the results for the same specimen with $t = 2 \times 10^{-2}$ sec, and

$$E_a(r,t) = 0.7 \times 10^{10}(32 \times 10^3 t)\exp(-5 \times 10^4 t)\exp(-0.11r^2) \qquad (8.2.12)$$

8.3 Temperature Distribution in Electroslag Welding

At this point we present the application of finite element simulation to the analysis of electroslag welding [Sluzalec, 1989]. A diagram of this process is presented in Fig. 8-14. A current is passed from a consumable electrode through a molten slag and molten metal pool. The 'Joule heating' in the slag causes the electrode to melt and the droplets thus formed pass through the slag and collect in the metal pool. The solidification of the pool causes the

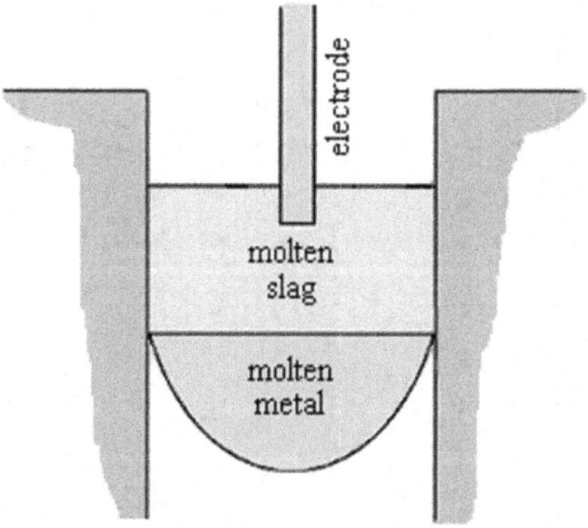

Figure 8-14. Scheme of electroslag welding

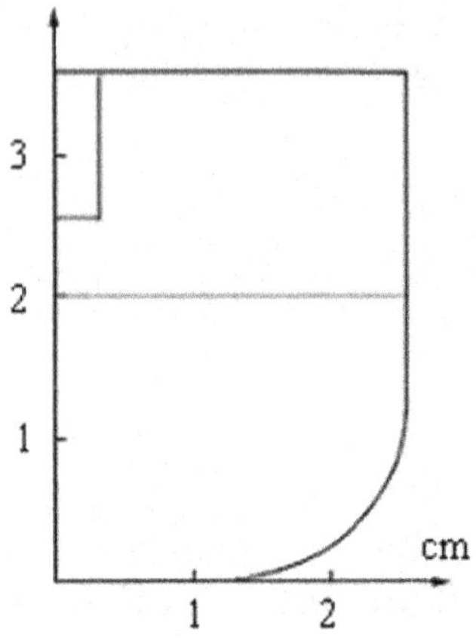

Figure 8-15. Region used to model electroslag welding

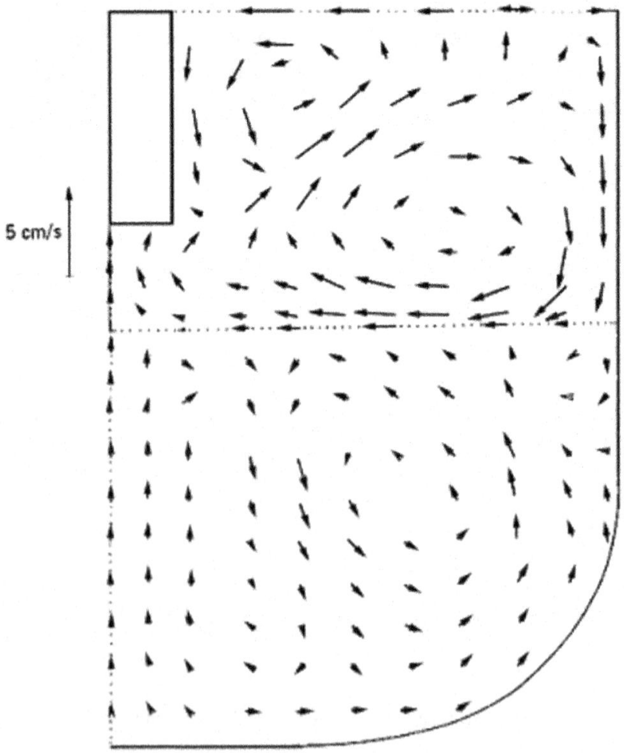

Figure 8-16. Velocity field for electroslag welding

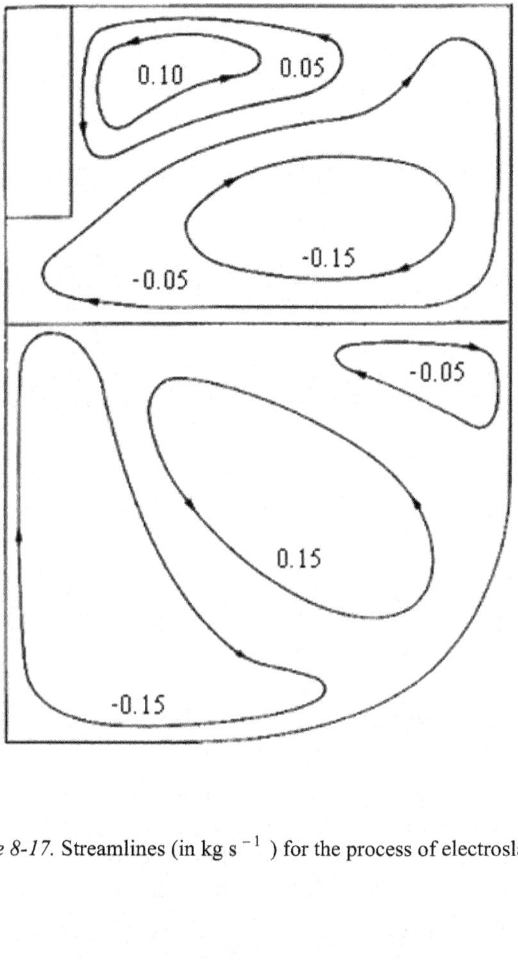

Figure 8-17. Streamlines (in kg s^{-1}) for the process of electroslag welding

Table 8-4. Thermal properties of molten slag and metal

	Specific heat ($kcal\ kg^{-1} K^{-1}$)	Thermal conductivity ($kcal\ m^{-1} s^{-1} K^{-1}$)	Density ($g\ cm^{-3}$)	Viscosity ($kg\ m^{-1} s^{-1}$)
Molten slag	0.2	0.0025	2.8	0.01
Molten metal	0.18	0.05	7.3	0.006

establishment of the joint connecting the two plates. Fluid motion in the system is represented by the fluid flow equations, which in essence express a balance between the rate of change of momentum within an infinitesimal fluid element and the sum of the net forces acting upon it. Fig. 8-15 presents the dimensions of the analyzed region. The results of the analysis are presented in Figs. 8-16 – 8-18. The material parameters assumed for the

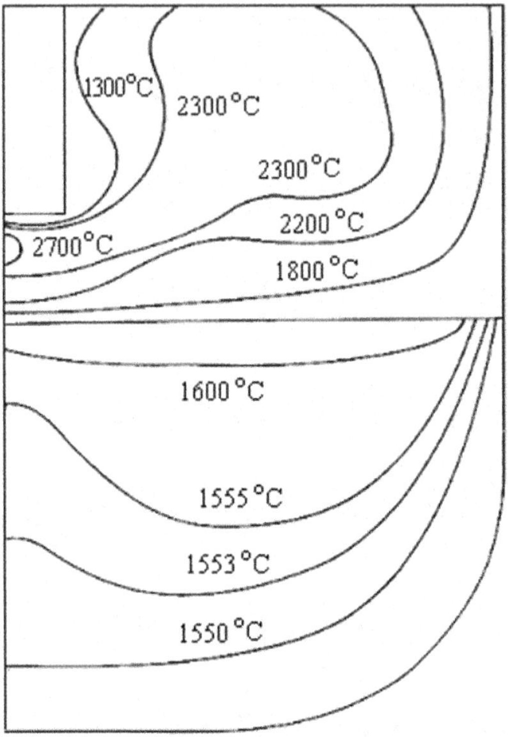

Figure 8-18. Computed temperatures for the process of electroslag welding

analysis are: melting temperature of the electrode 1553°C, melting temperature of the slag 1420°C. The specific heat of the molten slag and metal and other parameters are listed in Table 8-4.

Part IV

WELDING STRESSES AND DEFORMATIONS

Chapter 9

THERMAL STRESSES IN WELDING

9.1 Changes of Stresses During Welding

Changes of stresses during welding both in the weld and the base material are the result of welding heat source acting on the welded structure. This chapter analyses thermal stresses during welding. The changes of temperature and resulting stresses that occur during welding are presented in Fig.9-1.It is assumed that the weld is made along the axis x and the moving welding heat source is located at point *0*.

The mentioned figure shows the distribution of stresses along chosen cross sections. At cross-section 1-1 thermal stresses are almost zero. At cross-section 2-2 the stress near the welding heat source is equal to zero, because of existence of molten metal at this point.

Stresses in short distance from heat source are compressive, because the expansion of these areas is restrained by the surrounding metal where the temperatures are lower. Since the temperatures of these areas are high and the yield strength of the material low, stresses in these areas are as high as the yield strength of the material at corresponding temperatures.

The magnitude of compressive stress passes through a maximum increasing distance from the weld or with decreasing temperature. Stresses in regions away from the weld are tensile and balance with compressive stresses in areas near the weld.

At cross-section 3-3 the weld metal and base metal regions near the weld have cooled, they contract and cause tensile stresses in regions close to the weld. As the distance from the weld increases, the stresses first change to compressive and then become tensile.

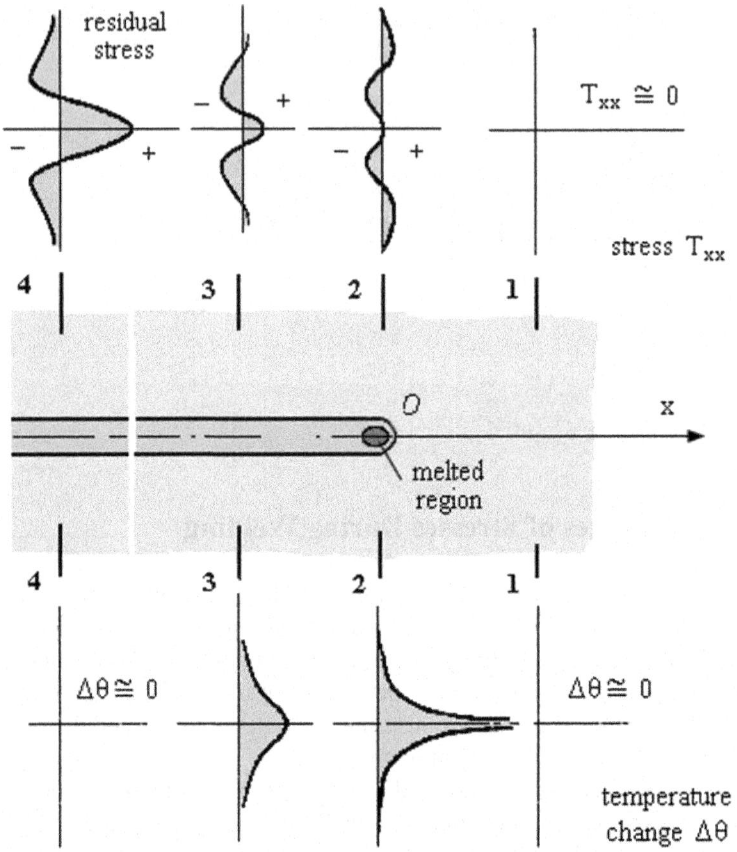

Figure 9-1. Changes of temperature and stresses during welding

At cross-section 4-4 high tensile stresses are produced in regions near the weld, while compressive stresses are produced in regions away from the weld. This is the usual distribution of residual stresses that remain after welding is completed. The region outside the cross-hatched area remains elastic during the entire welding thermal process.

9.2 Residual Stresses

9.2.1 Two-Dimensional Case

Consider 2D case welding stress analysis.The solution is obtained by means of finite element method. A plane strain idealization at mid-section is presented in Fig.9-2. The material properties assumed for computation are

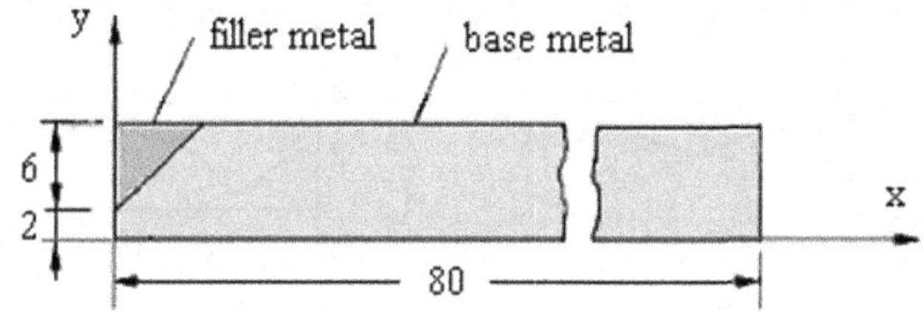

Figure 9-2. Weld under consideration

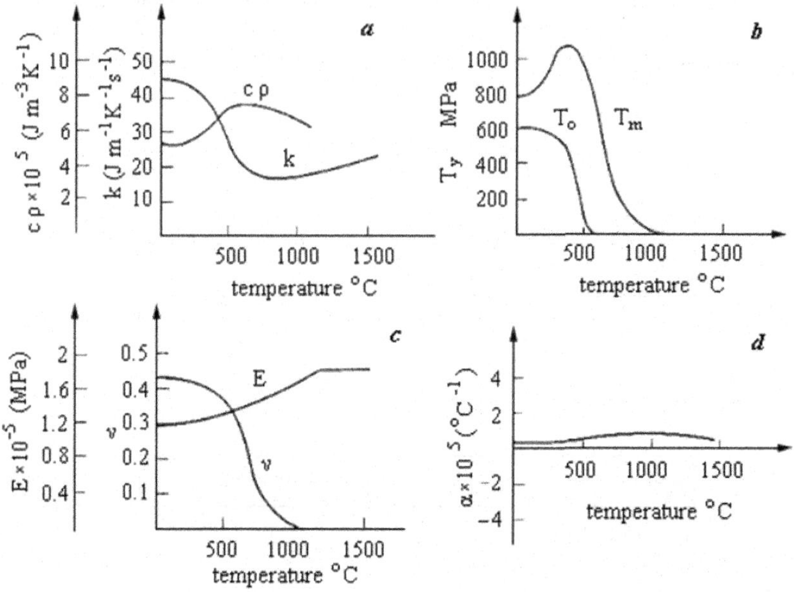

Figure 9-3. Properties of the base material assumed for the analysis: ***a*** thermal conductivity $k(\theta)$and heat capacity $c\ \rho(\theta)$; ***b*** initial $T_0(\theta)$and maximum yield limit $T_m(\theta)$; ***c*** elastic modulus $E(\theta)$and Poisson's ratio $\nu(\theta)$; ***d*** linear coefficient of thermal expansion $\alpha^T(\theta)$

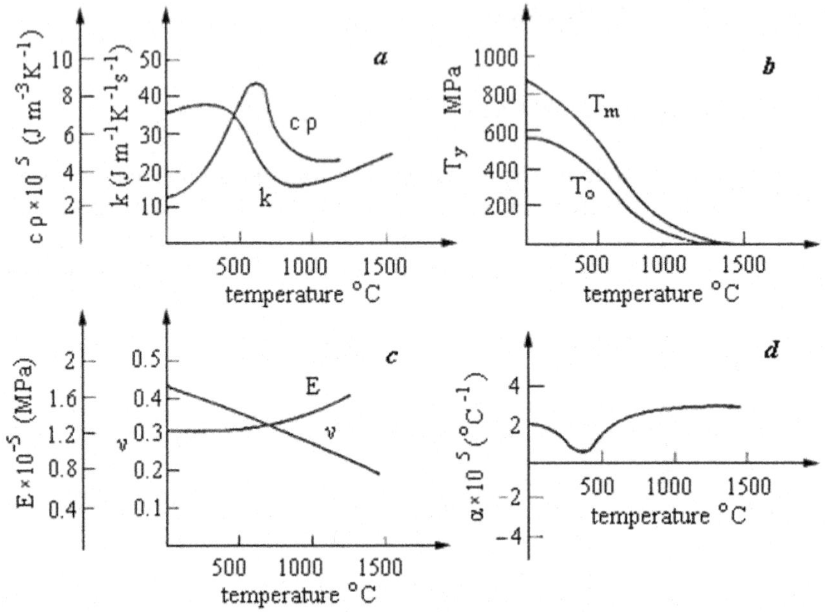

Figure 9-4. Properties of the filler metal assumed for the analysis: ***a*** Thermal conductivity $k(\theta)$ and heat capacity $c\,\rho(\theta)$; ***b*** initial $T_0(\theta)$ and maximum $T_m(\theta)$ yield limit; ***c*** elastic modulus $E(\theta)$ and Poisson's ratio $\nu(\theta)$; ***d*** linear coefficient of thermal expansion $\alpha^T(\theta)$

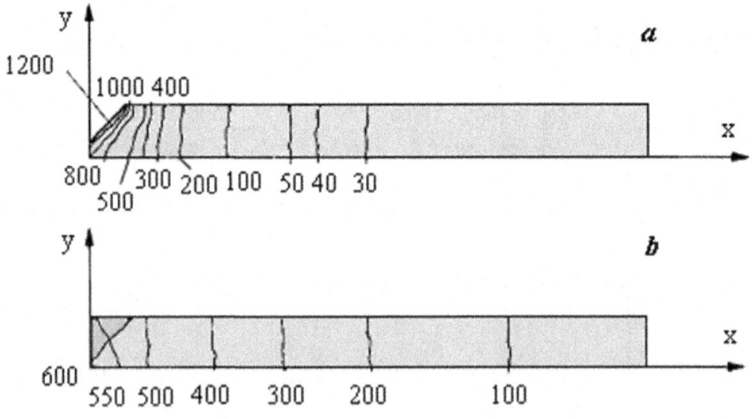

Figure 9-5. Temperature distribution in the weld (°C): ***a*** heating phase t = 5 seconds; ***b*** cooling phase t = 20 seconds

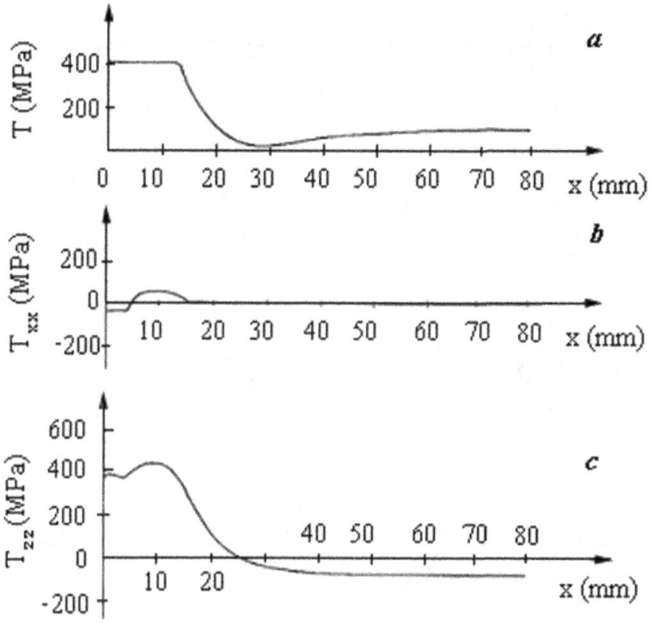

Figure 9-6. Residual stress distributions at the top of the mid-section: ***a*** equivalent Huber-Mises stress; ***b*** transverse welding stress, ***c*** longitudinal welding stress

given in Figs. 9-3 and 9-4. The distribution of temperature contours in the mid-section for heating and cooling phases for two chosen times of the process is given in Fig. 9-5. Residual stress distributions at the top of the mid-section are presented in Fig.96.

9.2.2 Plug Weld

Plug welded element made of low carbon steel is shown in Fig.9-7. The temperatures at the weld center is assumed to be 600°C. The distribution of residual stresses is shown in Fig. 9-8. The material properties used for computations are the same as in Section 921.

In the weld and adjacent areas tensile stresses equal to the yield stress of the material are observed both in radial and tangential directions.

In areas away from the weld, radial stresses T_r are tensile and tangential stresses T_θ are compressive. Both stresses decreased as the distance from the weld increased.

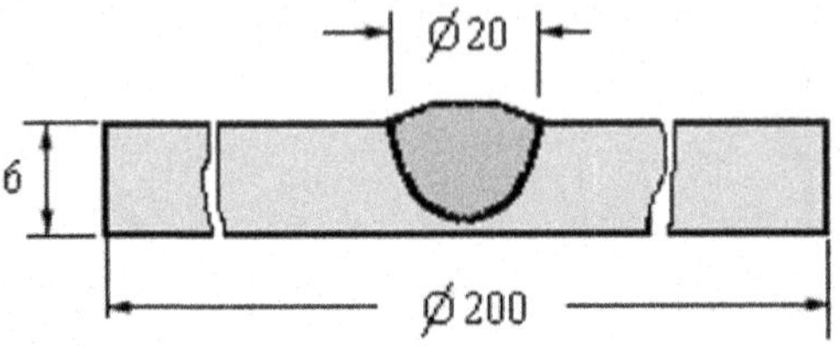

Figure 9-7. Plug-welded element

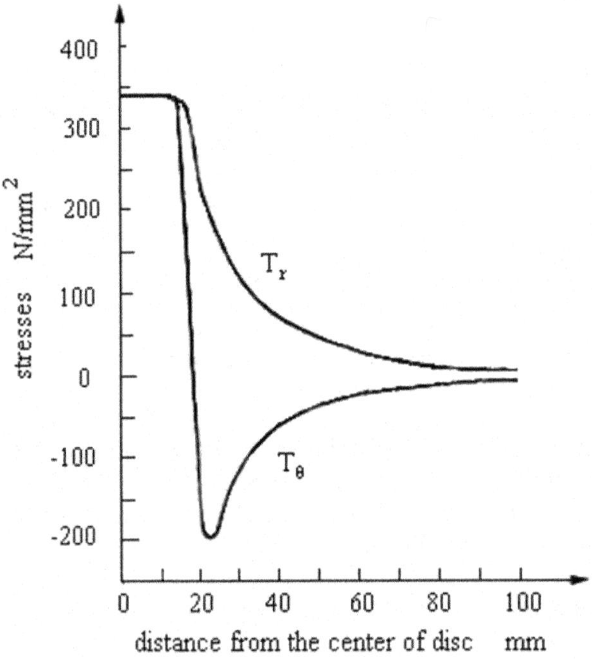

Figure 9-8. Distribution of residual stresses in a plug weld

9.2.3 Circular Patch Weld

Patch weld which are used in repair pbs is shown in Fig.9-9.Consider a circular plate welded into a large plate with a circular hole.Since shrinkage of the inner plate is restrained by the surrounding outer plate, high residual stresses are produced. The typical distribution of residual stresses in circular patch welds is shown in Fig. 9-9. The radial stresses T_r and tangential stresses T_θ are presented along the diameter. High tensile residual stresses exist in the weld area. The maximum of tangential stress is higher than the maximum radial stress. In the inner plate, radial and tangential stresses are tensile and approximately equal.

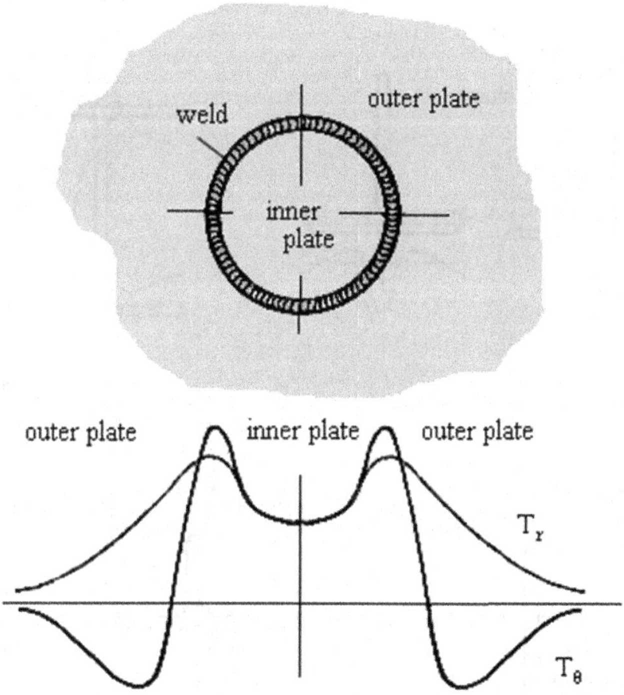

Figure 9-9. Residual stresses in a circular patch weld

Residual stresses in a path weld are produced primarily by shrinkage of the weld metal in the direction parallel to the weld or in the circumferential direction and shrinkage of the weld metal in the direction perpendicular to the weld or in the radial direction.

9.2.4 Welded Structures

Civil structures are often fabricated by welding. The typical distribution of residual stresses is shown in Fig.9-10.

First one shows residual stresses in a welded T-shape. High tensile residual stresses parallel to the axis are observed in areas near the weld in sections away from the end of the column. Stresses in the flange are tensile near the weld and compressive away from the weld. The tensile stresses near the upper edge of the weld are due to the longitudinal bending distortion caused by longitudinal shrinkage. Angular distortion is also observed.

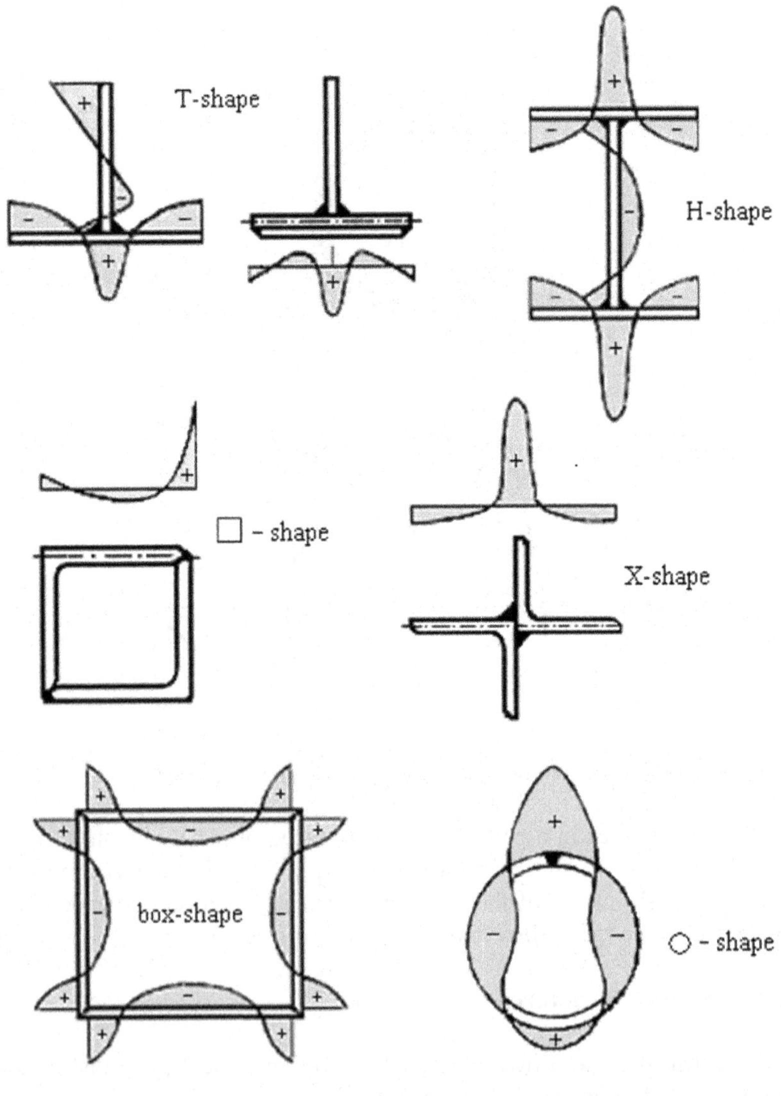

Figure 9-10. Residual stresses in welded shapes

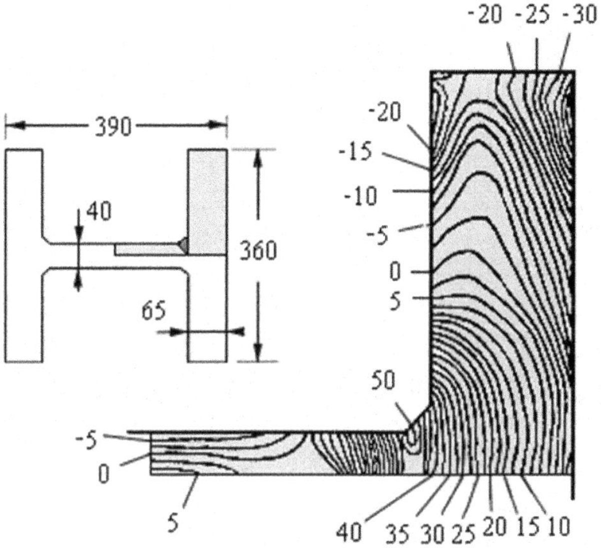

Figure 9-11. Variation of residual stresses (in KI)in a welded H-shape

The typical distribution of residual stress in an H-shape and a box shape are shown in Fig.9-10.

The residual stresses shown are parallel to the axis.These are tensile in areas near the welds and compressive in areas away from the weld.Fig.9-11 illustrates the calculated variation of residual stresses (in KI)in H-shape made of low-carbon steel with properties given in Section 921.

Chapter 10

WELDING DEFORMATIONS

10.1 Distortion

In the course of heating and cooling processes during welding thermal strains occur in the weld metal and base metal near the weld. The stresses resulting from thermomechanical loadings cause bending, buckling and rotation. These displacements are called distortion in weldments. In welding process three fundamental changes of the shape of welded structure are observed. These transverse shrinkage perpendicular to the weld line, longitudinal shrinkage parallel to the weld line and angular distortion (rotation around the weld line)These changes of welded structure are shown in Fig.10-1 where classification of distortion has been introduced.

The first figure on the left shows transverse shrinkage. It is shrinkage perpendicular to the weld line. The second figure on the top presents rotational distortion. It is an angular distortion in the plane of the plate due to thermal expansion. The figure in the middle illustrates angular change ie. transverse distortion. A non-uniform temperature field in the thickness direction causes distortion (angular)close to the weld line. The second figure in the middle shows longitudinal bending distortion. Distortion in a plane through the weld line and perpendicular to the plate. The first figure on the bottom illustrates longitudinal shrinkage, ie. shrinkage in the direction of the weld line. In the second figure on the bottom thermal compressive stresses cause instability when the plates are thin. It is called buckling distortion.

A typical structural component in ships, aerospace vehicles, and other structures is shown in Fig.10-2. It is a panel structure where a flat plate with longitudinal and transverse stiffness fillet is welded to the bottom. In the fabrication of panel structure distortion problem is caused by angular changes along the fillet welds.

The deflection of the panel changes in two directions as shown in Fig. 10-2 when longitudinal and transverse stiffeners are fillet-welded.

The typical distortion in two types of simple fillet-welded structures is shown in Fig.10-3.In both cases, the plates are narrow in one direction and the distortion can be considered as two-dimensional.

If a fillet-ϸint is free from external constraint then the structure bends at each ϸint and forms a polygon.If the stiffeners are welded to a rigid beam the angular changes at the fillet welds will cause a wavy, or arc-form distortion of the bottom plate.

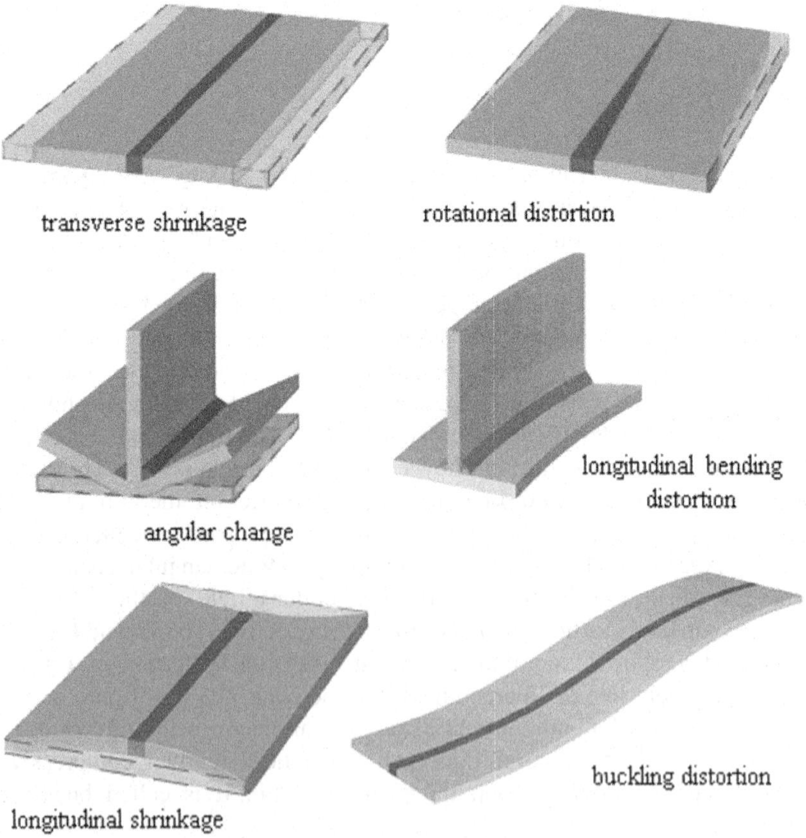

Figure 10-1. Types of weld distortion

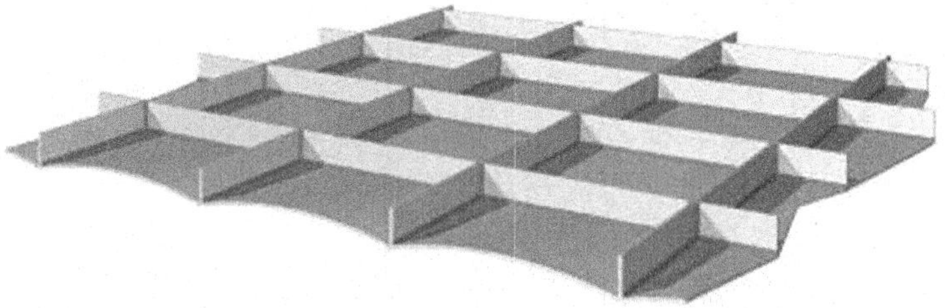

Figure 10-2. Welded structure with stiffeners

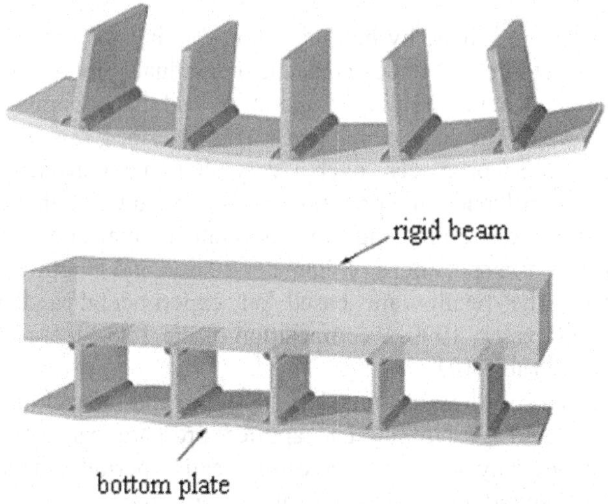

Figure 10-3. Types of distortion in fillet-welded structure

The above figures illustrates only the character of welding deformations. The amount of deformations should be calculated for each case of welding process and structure separately, because there not exist any simple formula for determination of welding deformations.

10.2 Deformations in Friction Welding

Friction welding is a complicated process, which involves the interaction of thermal and mechanical phenomena. In friction welding the heat for welding is obtained by conversion of mechanical energy to thermal energy at the interface of the workpieces.

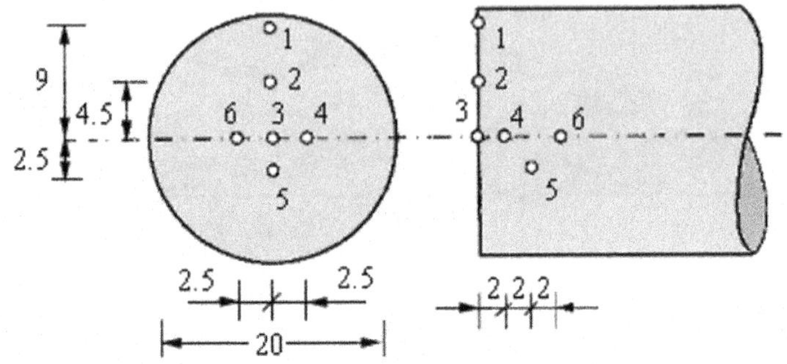

Figure 10-4. Workpiece with chosen points for temperature analysis

Friction welds are made by holding a non-rotating workpiece in contact with a rotating workpiece under constant or gradually increasing pressure until the interface reaches welding temperature, and then stopping rotation to complete the weld. The heat developed at the interface rapidly raises the temperature of the workpieces, over a very short axial distance. Welding occurs under the influence of a pressure that is applied while the heated zone is in the plastic temperature range. In this point thermal and displacement fields are analyzed in specimens of diameter 20 mm and length 75 mm made of 20G Steel. The results are based on experimental and numerical investigations [Sluzalec, 1990]A composition of the 20G Steel used in the study is given in Table 101.

The rotational speed was assumed as 1460 rpm and the welding pressures are 40 N/mm^2 and 60 N/mm^2. The applied pressure was held constant throughout the weld cycle, when the desired heating period had elapsed. The specimen for temperature measurements is presented in Fig. 10-4 with indicated points denoted by 1 through 6 where temperatures are monitored.

At the place of abutment the rate density of heat supply heat source is given by the following equation

$$q - \int_a \sigma \mu \omega r da \qquad (1021)$$

where σ is the axial pressure at the place of contact, μ is the coefficient of friction, ω is the angular speed, r is the radius and a is the surface upon which the heat rate acts. The following boundary conditions are used in the model: in the place of contact, as it is indicated above, the heat rate given by Eq.(1021) is assumed. The surface boundary conditions, eg. radiation and convection boundary conditions are introduced between the rod and the surrounding air. Surface forces are applied to simulate mechanical loading.

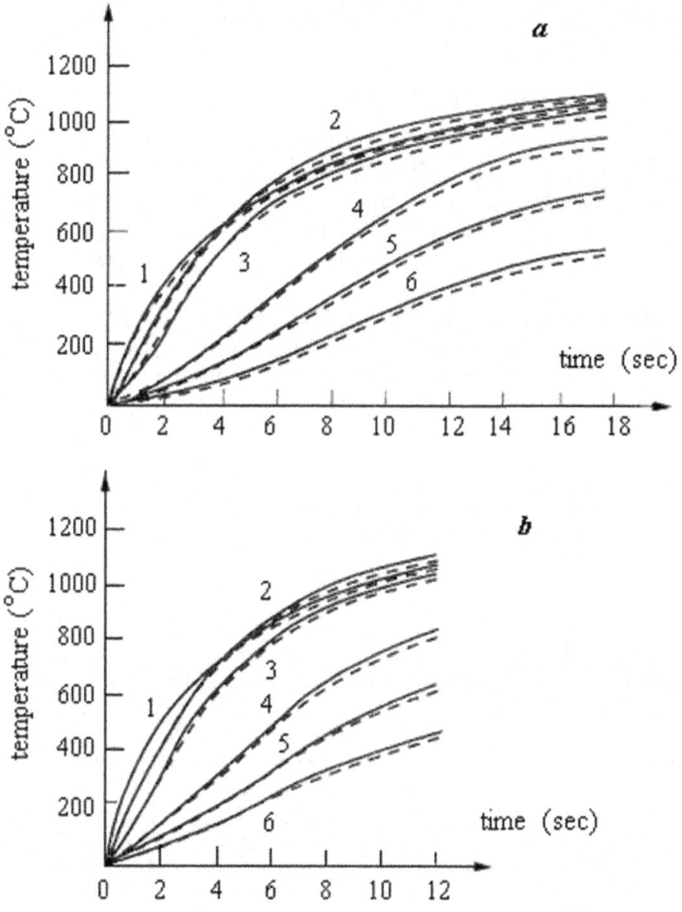

Figure 10-5. Heating cycles during friction welding for various welding pressures
a - 40 N/mm ^{2}, ***b*** - 60 N/mm ^{2}

These forces exists at the plane of contact between two elements which are welded. If a point moves out from the contact plane during the process the value of applied force is equal to zero.

Figure 10-5 shows the thermal cycles of welding at different points on a specimen. The temperatures calculated are denoted by continuous lines, results from measurements by dotted lines. With all the values of welding pressures investigated (for constant angular speed) the equalization of temperatures over the end occurred at quite an early stage of the process before the steady temperature had been reached.

Table 10-1. Composition of the 20G Steel

C	Si	Mn	Ni	Cr	P	S
025	035	1	-	-	005	005

Table 10-2. Material properties of the 20G Steel

Temperature (°C)	0	300	600	1000	1200
Thermal conductivity (W/m K)	147	152	198	253	281
Specific heat J/kg K	390	510	980	1300	680
Density (kg/m ^{3})	8 x 10 ^{3}	8 x 10^3	8 x 10^3	8 x 10^3	8 x 10^3
Surface film conductance (W/m ^{2} K)	009	009	009	009	009
Stefan-Boltzman constant (W/m ^{2} K^4)	577	577	577	577	577
Initial yield limit (MPa)	560	430	310	100	50
Friction factor	13	18	1	06	05

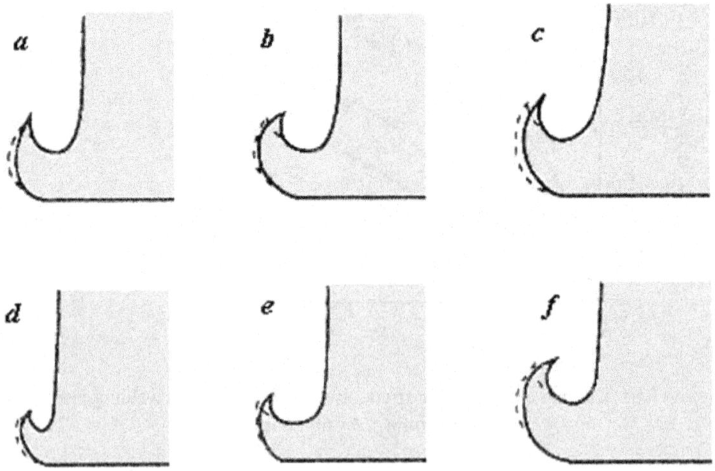

Figure 10-6. Real (–) and theoretical (-----) deformations of rods in friction welding for various parameters; heating time: (*a*) 7 s, (*b*) 12 s, (*c*) 18 s, (*d*) 3 s, (*e*) 7 s, (*f*) 12 s. Welding pressure: (*a* - *c*) 40 N/mm ^{2}, (*d* - *f*) 60 N/mm ^{2}

The friction factor as a function of temperature is given in Table 10-2. The majority of research workers believe that this relationship is dependent largely on the temperature conditions of the friction surfaces. Variations in

welding pressures primarily affect the rate of deformation of the rubbing surfaces and the temperature gradient. The temperature in turn has a substantial effect on the mechanical properties of the material. It is suggested that the limiting steady temperature in the joint cannot be higher than the temperature at which the yield point of the material is equal to the pressure used in the experiment. The maximum temperature in the joint during frictional heating depends not only on the pressure, but also on the temperature gradient which depends on the rotational speed in particular.
Figure 10-6 compares numerical results with experimental observations as to the shape of upset metal. Irregularities in the shape of the extruded material in a circumferential direction are not analyzed. These irregularities do not appear if specimens undergoing pressure during friction welding have exactly the same axis of symmetry and parallel surfaces of contact. Non-uniform properties of the material also can lead to irregularities in these shapes.

References and Further Reading

Adams, C. M., Jr., Cooling rate and peak temperatures in fusion welding, *Welding Journal*, **37** (5), Research Supplement, 210-215 (1958).

Alpsten, G. A., Tall, L., Residual stresses in heavy welded shapes, *Welding Journal*, **39** (3), Research Supplement, 93-105 (1970).

Anderson, J. E., Jackson J. E., Theory and application of pulsed laser welding, *Welding Journal*, **12**, (1965).

Apps, R. L., Milner D. R., Heat flow in argon-arc welding, *Brit. Weld. Journal*, 475-485 (October 1955).

Bentley, K. P., Greenwood, J. A., McKnowlson, P., Bakes, R. G., Temperature distributions in spot welds, *Brit. Weld. Journal*, 613-619 (1963).

Brown, S., Song, H., Finite element simulation of welding of large structures, *ASME Journal of Engineering for Industry*, **114**, (4), 441-451 (1992).

Brown, S., Song, H., Implications of three-dimensional numerical simulations of welding of large structures, *Welding Journal*, **71**, 55-62 (1992).

Brust, F. W., Dong, P., Zhang, J., A constitutive model for welding process simulation using finite element methods, Advances in Computational Engineering Science, *S. N. Atluri & G. Yagawa, Eds.*, 51-56 (1997).

Brust, F. W., Kanninen, M. F., Analysis of residual stresses in girth welded type 304-stainless pipes, *ASME Journal of Materials in Energy Systems*, **3** (3) (1981).

Carslaw, H. S., Jaeger J. C., Conduction of heat in Solids, *Oxford University Press*, (1959).

Christensen, N., Davis, V. de L., Giermundsen, K., Distribution of temperatures in arc welding, *British Welding Journal*, **12** (2), 54-75 (1965).

Dilthey, U., Habedank, G., Reichel, T., Sudnik, W., Ivanov, W. A., Numeric simulation of the metal-arc active gas welding process, *Welding & Cutting*, **45**, E50-E53 (1993).

Doty, W. D., Childs, W. J., A summary of the spot welding of high-tensile carbon and low-alloy steels, *Welding Journal*, **25** (10), Research Supplement, 625-630 (1946).

Eager, T. W., Tsai N. S., Temperature fields produced by traveling distributed heat sources, *Welding Journal*, **62** (12), 346-355 (1983).

Earvolino, L. P., Kennedy, J. R., Laser welding of aerospace structural alloys, *Welding Journal*, **3**, (1966).

Flugge, W., Four-place tables of transcendental functions, *McGraw-Hill*, New York, N. Y. (1954).

Goldak, J., Chakravarti, A., Bibby, M., A new finite model for welding heat sources, *Metallurgical Transaction*, 15B, 299-305 (1984).

Goldenberg, B. H., A problem in radial heat flow, *Brit. Int. Appl. Phys.*, **2**, 233-237 (1951).

Goldsmith, A., Waterman, T. E., Hisschhorn, H. J., Handbook of thermo-physical properties of solid materials, Vol. I and Vol. II, *Pergamon Press*, (1961)-(1963).

Greenwood, J. A., Temperature in spot welding, *Brit. Weld. Journal*, 316-322 (1961).

Grosh, R. J., Trabant, E. A., Arc-welding temperatures, *Welding Journal*, **35** (8), Research Supplement, 396-400 (1956). Cheng, C. J., Transient temperature during friction welding of similar materials in tubular form, *Welding Journal*, **41** (12), Research Supplement, 542-550 (1962).

Grosh, R. J., Trabant, E. A., Hawkins, G. A., Temperature distribution in solids of variable thermal properties heated by moving heat sources, *Quart. Appl. Math.*, **13** (2), 161-167 (1955).

Hess, W. F., Merrill, L. L., Nippes, E. F., Jr., Bunk, A. P., The measurement of cooling rates associated with arc welding and their application to the selection of optimum welding conditions, *Welding Journal*, **22** (9), Research Supplement, 377-422 (1943).

Jackson, C. E., Shrubsall, A. E., Energy distribution in electric welding, *Welding Journal*, **29** (5), Research Suppl., 232-242 (1950).

Jeong,S. K., Cho, H. S., An analytical solution to predict the transient temperature distribution in fillet arc welds, *Welding Journal*, **76** (6), 223-232 (1997).

Jhaveri Pravin Moffatt, W. G., Adams, C. M., Jr., The effect of plate thickness and radiation on heat flow in welding and cutting, *Welding Journal*, **41** (1), Research Supplement, 12-16 (1962).

Karman, T. V., Biot, M. A., Mathematical methods in engineering, *McGraw-Hill*, New York, N. Y. (1940).

Kasuya, T., Yurioka, N., Prediction of welding thermal history by a comprehensive solution, *Welding Journal*, **72** (3), 107-115 (1993).

Kawai, T., A study on residual stresses and distortion in welded structures, *Journal of the Japan Welding Society*, **33** (3), 314 (1964).

Kawai, T., Yoshimura, N., A study on residual stresses and distortion in welded structures (Part 2), *Journal of the Japan Welding Society*, **34** (2), 214 (1965), and Part 3, **34** (12), 215 (1965).

Lancaster, J. F., Energy distribution in argon-shielded welding arcs, *Brit. Weld. Journal*, **1**, 412-426 (1954).

Lesnewich, A., Control of melting rate and metal transfer in gas-shielded metal-arc welding: Part I, Control of electrode melting rate, *Welding Journal*, **37** (8), Research Supplement, 343-353 (1958); Part II: Control of metal transfer, *Welding Journal*, **37** (9), 418-425 (1958).

Murray J. D., Welding of high yield point steels, *Welding and Metal Fabrication*, **8**, (1966).

Myers, P. S., Uyehara, O. A., Borman, G. L., Fundamentals of heat flow in welding, *Welding Research Council Bulletin*, 123 (July 1967).

Nagaraja Rao, N. R., Esatuar, F. R., Tall, L., Residual stresses in welded shapes, *Welding Journal*, **43** (7), Research Supplement, 295-306 (1964).

Nguyen, N. T., Ohta, A., Matsuoka, K., Suzuki,N., Maeda, Y., Analytical solutions for transient temperature of semi-infinite body subjected to 3-D moving heat sources, *Welding Journal*, **78** (8) (1999).

Nippes, E. F., Merrill, L. L., Savage, W. F., Cooling rates in arc welds in ½ inch plate, *Welding Journal*, **28** (11), Research Supplement, 556-563 (1949).

Prokhorov, N. N., Samotokhin, S. S., Effect of artifical flowing off of Heat on processes of developing internal stresses and strain in welding, *Avt. Proiz.*, **5**, 63-69 (1977).

Pugin, A. I., Pertsovskii, G. A., Calculation of the thermal cycle in the HAZ when welding very thick steel by the electroslag process, *Avt. Svarka*, **6**, 14-23 (1963).

Rabkin, D. M., Temperature distribution through the weld pool in the automatic welding of aluminium, *British Welding Journal*, **6** (8), 132-137 (1959).

Ramsey, P. W., Chyle J. J., Kuhr J. N., Myers, P. S., Weiss, M., Groth, W., Infra-red temperature sensing for automatic fusion welding, *Welding Journal*, **42** (8), Research Supplement, 337-346 (1963).

Roberts, Doris K., Wells, A. A., Fusion welding of aluminum alloys, Part V, A mathematical examination of the effect of bounding planes on the temperature distribution due to welding, *Brit. Weld. Journal*, 553-560 (December 1954).

Robinson, M. H., Observations on electrode melting rates during submerged-arc welding, *Welding Journal*, **40** (11), Research Supplement, 503-515 (1961).

Rosenthal, D., Cambridge, M., The theory of moving source of heat and its application to metal treatments, *Trans. ASME*, **68** (11), 849-866 (1946).

Rosenthal, D., Mathematical theory of heat distribution during welding and cutting, *Welding Journal*, **20** (5), Research Supplement, 220-234 (1941).

Rosenthal, D., The theory of moving sources of heat and its application to metal treatments, *ASME Trans.*, 849-866, (1946).

Rykalin, N. N., Berechnung der Wärmevorgänge beim Schweissen, *Verlag Technik*, Berlin, 68-69 (1957).

Rykalin, N. N., Calculation of heat processes in welding, *Lecture the presented before the American Welding Society*, (April 1961).

Rykalin, N. N., Calculations of thermal processes in welding, *Mashgiz*, Moscow (1951).

Służalec, A., A new finite element approach to heat flow analysis in 3D developable structures, *Journal of Thermal Analysis*, **30**, 1063-1069 (1985).

Służalec, A., An analysis of thermal phenomena in electromagnetic field during electroslag welding, *Int. J. Computers & Fluids*, **17** (2), 411-418 (1989).

Służalec, A., An analysis of titanium weld failure, *Proceed. Int. Conf. JOM-4*, Denmark, (1989).

Służalec, A., An evaluation of the internal dissipation factor in coupled thermo-plasticity, *Int. J. Nonlinear Mechanics*, **25** (4), 395-403 (1990).

Służalec, A., An influence of laser beam intensity on the flow of metal, *Proceeding Int. Conf. Laser Materials Processing for Industry, ed. S.K.Ghosh*, LASER-5, (1989).

Służalec, A., Bruhns, O. T., Thermal effects in thermo-plastic metal with internal variables, *Computers & Structures*, **33** (6), 1459-1464 (1989).

Służalec, A., Computational methods for diffusion problems in welded joints, *Proceedings International Conference JOM-2*, Denmark, 322-327 (1984).

Służalec, A., Computer analysis of melting processes in metals undergoing laser irradiation, *Proceeding Int. Conf. High Power Lasers in Metal Processing, ed. S.K.Ghosh and T.Ericsson,* LASER-4, (1988).

Służalec, A., Flow of metal undergoing laser irradiation, *Numerical Heat Transfer*, 253-263, (1988).

Służalec, A., Grzywiński, M., Stochastic convective heat transfer equations in finite differences method, *Int. J. of Heat & Mass Transfer*, **43**, 4003-4008 (2000).

Służalec, A., Kleiber, M., Finite element analysis of heat flow in friction welding, *Eng. Transac.*, **32**, (1), 107-113 (1984).

Służalec, A., Kleiber, M., Numerical analysis of heat flow in flash welding, *Arch. Mech.*, **35**, (5-6), 687-699 (1983).

Służalec, A., Kubicki, K., Sensitivity analysis in thermo-elastic-plastic problems, *J. Thermal Stresses*, **25**, 705-718 (2002).

Służalec, A., Kysiak, A., An analysis of weld geometry in creep of welded tube undergoing internal pressure, *Computer & Structures*, **40** (4), 931-938 (1991).

Służalec, A., Kysiak, A., Creep of welded tubes undergoing internal pressure, *Archive. of Mechanical Engineering* (in Polish), **3**, 169-179 (1992).

Służalec, A., Moving heat source analyzed by finite elements and its application in welding, *Indian Journal of Technology*, **24**, 303-308 (1986).

Służalec, A., Random heat flow with phase change, *Int. J. of Heat & Mass Transfer*, **43**, 2303-2312 (2000).

Służalec, A., Shape optimization of weld surface, *Int. J. Solids & Structures*, **25** (1), 23-31 (1989).

Służalec, A., Simulation of the friction welding process, *ed. A.Niku-Lari, SAS World Conference*, IITI-International, 333-340 (1988).

Służalec, A., Solution of thermal problems in friction welding - comparative study, *Int. J. Heat & Mass Transfer*, **36** (6) 1583 - 1587 (1993).

Służalec, A., Temperature field in random conditions, *Int. J. Heat & Mass Transfer*, **34** (1), 55-58 (1991).

Służalec, A., Temperature rise in elastic-plastic metal, *Comp. Meth. Appl. Mech. Eng.* **96**, 293-302 (1992).

Służalec, A., Thermal effects in laser microwelding, *Computers & Structures*, **25** (1), 29-34 (1987).

Służalec, A., Thermal effects in friction welding, *Int. J. Mech. Sci.*, **32** (6), 467-478 (1990).

Służalec, A., Thermoplastic effects in friction welding analyzed by finite element method, *Proceedings International Conference on the Effects of Fabrication Related Stresses, Ed. The Welding Institute*, England, Paper 8, P8-1 – P8 -8 (1985).

Stoeckinger, G. R., Calabrese, R. A., Menaul, R. F., Computerized prediction of heat distribution in weld tooling, *Welding Journal*, **49** (1 and 6), Research Supplement, 14-26 and 272-277 (1970).

Sudnik, W., Ivanov, A., Mathematical model of the heat source in GMAW. Part I. Normal process, *Welding International*, **13**, 215-221 (1999).

Sudnik, W., Radaj, D., Erofeew, W., Computerized simulation of laser beam welding, model and verification, *J. Phys. D: Appl. Phys.*, **29**, 2811-2817 (1996).

Sudnik, W., Research into fusion welding technologies based on physical-mathematical models, *Welding & Cutting*, **43**, E216-E217 (1991).

Sutherland, J. D., High yield strength waterquenched and tempered steel plate, *Welding and Metal Fabrication*, **1**, (1967).

The Physical Properties of a Series of Steels, Part II, *Jnl. Iron and Steel Inst.*, **154** (2), 83-121 (1946).

Veron, C. W. J., Ultrasonic welding, *Welding and Metal Fabrication*, **11**, (1965).

Vogel, L. E., Lyens, J. V., Pumphrey, W. I., Temperature and hardness distribution in welded Al-4% Cu alloy sheet, *Brit. Weld. Journal*, 252-259 (June 1954).

Wayman, S. M., Stout R. D., A study of factors effecting the strength and ductility of weld metal, *Welding Journal*, **5**, (1958).

Weiss. S., Ramsey, J. N., Udin, H., Evaluation of weld-cracking tests on armor steel, *Welding Journal*, **35** (7), Research Supplement, 348-356 (1956).

Wells, A. A., Heat flow in welding, *Welding Journal*, **31** (5), Research Supplement, 263-267 (1952).

Wells, A. A., Oxygen cutting, *Brit. Weld. Journal,* 86-92 (1961).

Wilson, J. L., Claussen, G. E., Jackson, C. E., The effect of I^2Rheating on electrode melting rate, *Welding Journal,* **35** (1), Research Supplement, 1-8 (1956).

Wilson, W. M., Hao, C. C., Residual stresses in welded structures, *The Welding Journal*, **26** (5), Research Supplement, 295-320 (1974).

Yang, Y. P., Brust, F. W., Welding-induced distortion control techniques in heavy industries, Symposium on Weld Residual Stresses and Fracture 2000, ASME Pressure Vessels and Piping Conference, Seatle, WA, USA, July 23-27 (2000).

Yoshida, T., Abe, T., Onoue, H., Residual stresses in circular-patch-welds, *Journal of the Society of Naval Architects of Japan*, **105** (1959).

Subject Index